Democratic Reason

Democratic Reason

POLITICS, COLLECTIVE INTELLIGENCE,
AND THE RULE OF THE MANY

Hélène Landemore

PRINCETON UNIVERSITY PRESS

Princeton and Oxford

Copyright © 2013 by Princeton University Press
Published by Princeton University Press, 41 William Street, Princeton, New Jersey 08540
In the United Kingdom: Princeton University Press, 6 Oxford Street, Woodstock, Oxford-shire OX20 1TR

press.princeton.edu

Cover design by Marcella Engel Roberts. Cover photograph: 3-D rendering of a room with a circle of nine white chairs. © Franck Boston. Courtesy of Shutterstock.

First paperback printing, 2017

Paper ISBN: 978-0-691-17639-0

The Library of Congress has cataloged the cloth edition as follows:

Landemore, Hélène, 1976–
    Democratic reason : politics, collective intelligence, and the rule of the many / Hélène Landemore.
        p. cm.
    Includes bibliographical references and index.
    ISBN 978-0-691-15565-4 (hardcover : alk. paper) 1. Democracy. 2. Democracy—Philosophy. 3. Majorities. I. Title.
    JC423.L3355 2012
    321.8—dc23

                                                                    2012020314

British Library Cataloging-in-Publication Data is available

This book has been composed in Sabon

Printed on acid-free paper. ∞

Printed in the United States of America

To my parents

"The many, who are not as individuals excellent men, nevertheless can, when they have come together, be better than the few best people, not individually but collectively, just as feasts to which many contribute are better than feasts provided at one person's expense."

—Aristotle

"Democracy is the recurrent suspicion that more than half of the people are right more than half of the time."

—E. B. White

Contents

Acknowledgments

As BEFITS A BOOK ON the topic of collective intelligence, I should start by acknowledging that this book would not have been possible without many, many minds besides my own. I might not be able to do justice to all of them, but I hope that the resulting work does and, importantly, verifies, to a degree, its own thesis!

This book originates in a PhD dissertation conducted under the supervision of my advisors at Harvard University. Richard Tuck will probably not recognize his influence in the ideas advocated in it, and yet they benefited greatly from our many discussions. Jenny Mansbridge was a strong supporter of the project from the beginning, and it owes a lot to her insight and unfailing enthusiasm. I'm also grateful for Nancy Rosenblum's wisdom and sharp advice.

I owe a special debt to Jon Elster, my external advisor from Columbia University and informal mentor for many years, who encouraged me to think boldly and proved crucial in shaping the final version of the manuscript both through his direct comments and the common work we did at the Collège de France in Paris in 2007–2008, including the organization of a conference, now turned into a book, on collective wisdom.

My next greatest debt goes to Scott Page, whose work on the centrality of cognitive diversity in the phenomenon of collective intelligence was such a revelation and provided me with the missing argumentative link between the epistemic value of democratic procedures and their inclusiveness. I'm grateful for his insights and the time he took to reread some of the passages in this book.

From my time at Harvard University, I would also like to thank the scholars I interacted with at the Kennedy School's Center for Ethics, including Amalia Amaya, Michael Blake, Japa Pallikkathayil, Simon Rippon, and Annie Stilz; the members of the Project on Justice, Welfare, and Economics at the Weatherhead Center for International Affairs, particularly professors Amartya Sen, Thomas Scanlon, and Frank Michelman, as well as Quoc-Ahn Do (for one long discussion over coffee involving questions of threshold and maze metaphors); Kyoko Sato, then a colleague at the Center for European Studies; and the participants in the Harvard Graduate Political Theory Workshop, in particular professors Michael Frazer, Michael Rosen, and Eric Beerbohm, as well as Josh Cherniss, Sam Goldman, David Grewal, Susan Hamilton, Sean Ingham, Jo Kochanek, Matt Landauer, Lucas Stanczyck, and Elina Treyger.

From my year at the Collège de France in Paris, I would like to thank Pierre Rosanvallon, who invited me to present my work at his seminar and whose course on democracy at MIT a few years back had initially sparked my interest in the Condorcet Jury Theorem; and Florent Guénard for useful comments on this same presentation and a friendly collaboration on other projects.

From my semester at Brown University, I would like to thank the members of the Political Theory Project, specifically David Estlund, whose work on epistemic democracy is so foundational to mine, Sharon Krause, John Tomasi, and the group of young scholars I had the privilege to interact with there: Jason Brennan, Corey Brettschneider, Barbara Buckins, Yvonne Chiu, Emily Nacol, and Andrew Volmert.

I am now blessed with some wonderful colleagues at Yale University. Bryan Garsten, Karuna Mantena, Andrew March, and Paulina Ochoa, as well as Deme Kasimis (now at California State Long Beach) and Daniel Viehoff (from the University of Sheffield), who both joined our group that year, deserve special credit for spending time and brain power helping me improve a chapter that, unfortunately, did not make the cut (sorry!).

I am grateful to the jury of the Montreal Political Theory Workshop Award at McGill University, who recognized my work when it was still in search of a publisher. Arash Abizadeh, Pablo Gilabaert, Dominique Leydet, Jacob Levy, Catherine Lu, Christian Nadeau, Hasana Sharp, Stuart Soroka, Christina Tarnopolsky, Daniel Weinstock, Jürgen de Wispelaere, and the other participants in the workshop provided invaluable comments on an advance draft of this book and made it infinitely better.

Among the many other people who helped along the way, I'd like to thank Gerald Gaus, Gerry Mackie, Bernard Manin, John McCormick, Andrew Rehfeld, Keith Sutherland, and Nadia Urbinati. Thank you also to Rob Tempio and the team at Princeton University Press for their impressive professionalism. Thank you to the anonymous reviewer for that press for his endorsement and constructive criticisms. Finally, very warm thanks to Erin Pineda and Michael Turner for their help in editing the manuscript and tracking references.

A number of institutions supported my research over the last few years. At Harvard University, the Government Department; the Center for Ethics at the Kennedy School; the Project on Justice, Welfare, and Economics of the Weatherhead Center for International Affairs; and the Minda de Günzburg Center for European Studies offered me space and resources. The Collège de France, Brown University, and MIT provided welcome shelters for the two years before I found a home at Yale University.

I also want to thank the core of family and friends that have been with me, even at a distance, through the highs and lows of that academic adventure (and many nonacademic ones as well): my parents first and foremost, Gérard and Jacqueline, to whom this book is dedicated; my sister Marie and my brother Pierre; and a group of inspiring women that I'm lucky to call friends: Antonia, Hélène, Isabelle, Karine, Meredith, Sandra, and Susan; and finally, Darko, my partner in life and the father of my beautiful daughter. You came at the end of this long journey and made it all worth it.

Prologue

On May 29, 2005, France held a referendum on ratifying the European Union Constitution. Contrary to the hopes and expectations of the political and intellectual elites, the French people decided to reject this constitutional treaty. Commentators were all but unanimous in their condemnation of the result. Doubtless, they said, the people had gotten it all wrong; they had taken an ill-chosen opportunity to sanction the government (and Jacques Chirac, then the president of France, in particular), thereby irresponsibly sinking the European project.

As a French citizen, I initially reacted to the referendum result in similar fashion. How could my fellow citizens have voted in this ignorant way? If only they had looked past their navels, they would have realized that Europe represented the only possible future for a country having difficulties coming to terms with the economic and cultural challenges of globalization on its own. The proposed constitution was a perhaps imperfect but ultimately reasonable compromise between different political conceptions of the European project. In any case, embracing it was the only way to move the construction of Europe forward—a construction all the more necessary, as it represented the promise of modernization and necessary institutional reforms. There was a right answer to the referendum, and I believed the French people had given the wrong one.

Upon reconsideration, however, I came to think that as a single individual, with a particular and limited experience of what it is to be French in Europe today, I might not be in the best position to pass a judgment on the needs of France. Was it plausible that more than 15 million people should be wrong and I right? Arguably, I could find comfort in the fact that nearly 13 million of my fellow citizens voted with me. At the end of the day, however, was it more likely that 55 percent of the population was wrong and the other 45 percent right or vice versa? On some assumptions, this is not what the law of large numbers (as formalized in the Condorcet Jury Theorem, for example) would predict. At the very least, it seemed to me that some probabilistic consideration of this sort should have been factored into my assessment of the referendum results. I thus came to wonder whether there wasn't, after all, a good reason to trust the majority more than my own judgment and more than the minority of which I was a part. More broadly, I wondered, what if the reason we use majority rule in the first place is because it is in general a reliable decision procedure?

Immediate objections spring to mind, offering examples of deeply mistaken majorities. A majority put Socrates to death. A majority allegedly brought Hitler to power.[1] All over the world, in fact, formal and informal majorities endorse irrational, xenophobic, racist, anti-Semitic, and sexist ideologies. What if the French majority that rejected the constitution project was just as mistaken, if not as evil? The fact that a majority agrees on a position does not say much, the suspicion goes, about the intrinsic value of that position.[2]

Another objection runs even deeper, challenging the very notion that there could be a right or wrong answer to political questions. For some, the "right" answer to the referendum on the constitutional treaty is simply the one that is procedurally determined by its outcome. In that view, the endorsement of majoritarian decisions is just one of the rules of the democratic game that citizens implicitly underwrite when casting their vote. The point, however, is not to figure out any independently given "right" answer.

The first objection comes naturally to any observer of history, particularly the history of discrimination against minorities. Yet pointing at historical cases of democratic failure does not amount to a fully fledged refutation of the general validity of democracy as a collective decision-making rule; moreover, many supposedly classical examples of democratic failures, such as Athens's infamous Sicily expedition, can always be challenged.[3] Further, for every democratic failure, one may point out a

---

[1] This was Daniel Cohn-Bendit's postreferendum remark to a journalist who pointed out that the negative result of the referendum was, after all, the choice of the majority. The historical truth, as is turns out, is that a plurality, not a majority, brought Hitler to power (see Ermakoff 2008 for an account of what that vote looked like).

[2] Jacques Rupnik offers an alternative reading of the referendum results, which neither condemns the people as stupid nor accepts that they were right on that particular issue. According to him, the referendum in France had little chance to elicit the "collective wisdom" of the multitude, because referenda in France are plebiscites. If referenda were much more of a tradition, however, the way they are for example in Switzerland, the results would probably have been more meaningful. The Swiss are used to deciding by referendum, both at the federal and the cantonal level, issues such as "Should there be mandatory army service?" or "Should the highway cut across this ancient cemetery?" The system of the "initiative" also allows the Swiss people to have some control over the questions and the way they are phrased. The problem in France, Rupnik argues, is not that the people are stupid, but that they are not trusted enough. As a result, when they finally have an opportunity to voice their concerns, they do so in a contrarian way (personal communication, May 2006).

[3] See Finley's analysis of the Sicily expedition, which provides a more global critique of elitist objections to popular rule: "The familiar game of condemning Athens for not having lived up to some ideal of perfection is a stultifying approach. They made no fatal mistake, and that is good enough. The failure of the Sicilian expedition in 415–413 BC was a technical command failure in the field, not the consequences of either ignorance or inadequate planning at home. Any autocrat or any 'expert' politician could have made the same errors" (Finley 1985: 33). For Finley, occasional democratic mistakes do not count against the idea that democracy can be smart, nor can they serve as an argument for elitist theorists of democracy.

democratic success, or at least a comparatively worse failure of nondemocratic regimes. In the game of "Who went the most wrong in history?" it is unclear that democracies are on the losing end.[4] As Machiavelli made the point in the Discourses on Livy, the elitist tradition that holds the masses in contempt relies on a biased and methodologically flawed comparison between unruly mobs and the rare instances of wise and good-willed princes. Machiavelli suggests that when you stop comparing apples and oranges and compare people and princes that are both "shackled by laws" (Machiavelli 1996: 117), the evidence is much more favorable to the people and not as favorable to princes. Machiavelli, for his part, concluded from his own historical observations that people were "of better judgments" than princes (Machiavelli 1996: 115–19).[5] Contemporary political scientists have statistical methods and tools that can be applied to challenging and verifying that claim. Political theorists, meanwhile, can also attempt the comparison from an a priori perspective, comparing abstract models of different rules and their expected properties.

If anything, therefore, examples of democratic failures should invite further inquiry into the comparison, both theoretical and empirical, of democratic decision procedures with nondemocratic ones. When it comes to majority rule in particular, such examples should raise the probabilistic question (which can be asked from both a theoretical and empirical perspective) of when and where a majority is likely to be right, and how this should affect the authority of democratic decisions.

The second objection stems from a purely proceduralist understanding of democracy, according to which the value of democracy and its decisions is assessed only in terms of their procedural fairness. This objection denies, or at least avoids asserting, the existence of any objective or substantive criteria by which to assess a democratic decision. In that view, democratic decisions are good because they are procedurally fair, not because they yield outcomes that are in some sense "good." Such a purely proceduralist commitment to the value of democracy, however, runs against the idea that when we argue and deliberate about politics and when we ultimately vote at the end of such a deliberative process, what we hope for is the triumph of "the unforced force of the better argument," according to Habermas's beautiful and suggestive

---

[4] Notice that arguments stressing examples of evil majorities, while powerful and valid to some extent, also feature in the strategy of every brand of antidemocrat. As such, they often reek of what the philosopher Jacques Rancière (2005) calls "the hatred of democracy": a tendency to selectively look for examples of democratic failures while ignoring democratic successes or the comparative results of competing decision procedures.

[5] See chapter 3 of this book for a brief survey of Machiavelli's analysis of the wisdom of the multitude. See also the work of John McCormick (2011), with whose interpretation I am in complete agreement and which I rehearse in part here.

formula.[6] If no alternative was truly better, in some sense, than the others, one might wonder: Why would politicians bother campaigning, that is, resort to reasons and arguments and try to inform the people? Do they not do this in the explicit hope of helping them make more enlightened judgments about politics? Why mail citizens the 300-page volume of the constitution if not for them to make up their mind as to the "rightness" of this document for the European project? Even critics who routinely deplore the low level of information and knowledge of average voters confirm the idea that there is an epistemic component to the value of political decisions.

This reflection prompted me to question two things. The first is the justification of democracy as an intrinsically worthy regime, in virtue of the values embodied by its procedures (equality, justice, and so forth). A purely proceduralist justification seems to leave democratic governments and their decisions on shaky grounds. For democracy and democratic decisions to be fully justified, if not legitimate, it seems there has to be something more to it than the values it embodies. There has to be some kind of substantial merit and, I would argue, some kind of "intelligence" as well. In this respect, David Estlund's work on the epistemic dimension of democratic authority (1997, 2008) raised and answered some important questions. It left me, however, with more questions as to the actual epistemic performance that could be expected from democratic rule.

Second, assuming this idea of democratic intelligence is valid and relevant, I came to question the idea that such intelligence should be no more than the aggregation of the intelligence of individuals. The collective intelligence of the citizens—what I call more broadly "democratic reason"—might in fact be distinct from individual reason writ large. Psychology and cognitive sciences, including the science of animal behaviors, show that intelligence can be a property of groups as well as of individuals. The phenomenon of "emergent intelligence" characterizes societies of social animals such as ants and bees. Another relevant concept is that of "distributed intelligence"—which posits intelligence as spread across both the individual agents themselves (mind and body) and their environment (institutions, language, symbolic systems, and other "cognitive artifacts").

These new concepts of collective intelligence—as emergent or distributed—have had little influence in democratic theory. One reason probably has to do with the fact that political theorists, like philosophers, are by training suspicious of the idea that large groups can be smart. Another reason may have to do with the focus on individuals that comes

---

[6]The German original reads "der zwanglose Zwang des besseren Arguments" (Habermas 1981: 47).

with the notion of reason as autonomy, and undergirds political science's principled commitment to methodological individualism. From the point of view of political scientists (at least those influenced by rational choice theory), the relevant agency units are assumed to be individuals and should not be located at the supra level of groups (or, for that matter, at the infra level of genes). Notions of collective or distributed intelligence may seem by contrast to raise the specter of social holism. Yet, far from it, it can be argued that these notions lend themselves to explanations in terms of individual choices and actions, in the same way that collective-action problems can be accounted for by the analytical tools and individualist methodology of social choice theory.

At the same time that my ideas were thus turning to an epistemic approach to democracy, the literature on "the wisdom of crowds" was becoming mainstream (the landmark being Surowiecki 2004), focusing on the then-not-so-well-understood phenomena of the predictive accuracy of information markets or the birth and almost overnight success of Wikipedia, the free online encyclopedia written cooperatively by nonexperts. In parallel, debates about the importance of the Internet, the possibility of cyber- or e-democracy, and the role of blogs and amateur citizen-journalists in the new informational sphere also revolved around the notion of collective intelligence and were sometimes explicitly linked, however loosely, to the idea and ideal of democracy (see most recently Coleman and Blumler 2009 and, for a revolutionary notion of "Wiki government," Noveck 2009). How this literature can be reconciled with the more classical paradigm of deliberative democracy, which emphasizes exchange of arguments over mere aggregation of dispersed information and knowledge, is one of the obvious and most interesting challenges that this book aims to address.

Finally, the political landscape itself became shaped, on both sides of the Atlantic, by ideas related to the concept of collective intelligence, in both its deliberative and its aggregative dimension. For the 2007 presidential elections in France, the Socialist candidate Ségolène Royal campaigned on the Deweyan theme of "citizens' expertise." Her argument was that in a complex and informed world, every citizen holds a parcel of the truth and that the best source of enlightened political decision is to be found not in experts, or even in professional politicians, but in the people themselves.[7] She thus advocated more direct forms of participatory

---

[7] Translated and adapted from an introductory text available on Royal's website, http://www.desirsdavenir.org/index.php?c=bienvenue, accessed March 10, 2007. Similar ideas are present in her current defense of "popular and participatory universities," which are gatherings meant to combine the expertise of scientists, political militants, and regular citizens (see "Descriptif: Les Universités Populaires et Participatives," accessed April 8, 2012, http://www.desirsdavenir.org/les-universites-populaires-et-participatives/descriptif.html).

democracy and presented herself less as a leader of the people and more as a recipient and a catalyst for their own judgment. Simmering in Royal's discourse was the notion of collective intelligence—the idea that political solutions are often best figured out by the people as a whole, when individuals talk to each other and contribute their bits of knowledge to a general public discussion.[8]

In the United States, and in a more aggregative than deliberative vein, the Obama administration started experimenting with new tools and techniques of "crowdsourcing." In 2008, for example, it encouraged the creation of a "peer-to-patent" experiment, in which the process of patent reviewing was opened up from a few experts to a larger public. More significant perhaps was the launch in late May 2009 of the Open Government Initiative.[9] The principles of this initiative—transparency, participation, and collaboration—were explicitly about using the latent collective intelligence distributed across the nation to bring to the fore new policy ideas. While the Obama campaign was praised for its innovative way of reaching the hearts and wallets of average citizens, as opposed to the usual interest groups and big donors, later developments showed that the new administration aimed to go beyond culling votes and money and was seeking out knowledge, information, and ideas from usually voiceless citizens.[10]

[8]This approach to politics was, on the face of it, revolutionary. According to a commentator, the logic of "ségolisme" contributed to displacing the traditional distinction between those who know (the professional politicians) and those who don't (average citizens), as well as to integrating the life and problems of ordinary citizens into the noble sphere of "the political." Zaiki Laïdi, "Le véritable apport de Mme Royal," Le Monde, July 4, 2006. It is also ironic that someone literally named "kingly" (the meaning of "royal" in French) and who is a pure product of one of the most elitist French institutions (Ecole Nationale d'Administration) should claim the superiority of the people's judgments over that of individuals like herself. In any case, this new trend in politics is echoed everywhere, from the writing of encyclopedias on the Internet (e.g., Wikipedia) to the rise of a new form of journalism more popular and participatory than the traditional ones. Rue89 in France, for example, is an online newspaper started by former journalists from Libération that builds in part on the (controlled) input of nonjournalists. No longer passive readers, and not professional journalists either, the participants in these new forms of information processing contribute to revolutionizing the information industry along what are arguably participatory and democratic lines.

[9]Perhaps following the French example, at least if one is to believe Ségolène Royal, who claims that Obama was inspired by her campaign and copied it, borrowing in particular the idea of the citizen expert and adapting her tactic of participative democracy to the American political landscape (Le Monde, January 29, 2008). Given that Royal herself borrowed her ideas about participatory democracy from the American philosopher John Dewey, this seems only fair (if true). Another distinct and probably more direct influence is probably that of Cass Sunstein (himself influenced by the aggregative rather than deliberative preferences of James Surowiecki).

[10]See http://www.whitehouse.gov/open/, accessed June 10, 2009.

The ideas of wisdom of the crowds and the collective intelligence of regular citizens have thus been slowly gaining ground. Among the boldest implementations, Iceland's recent experiment of crowdsourcing the very writing of its new constitution is worth mentioning. Following the institutional crisis that resulted from the massive financial and economic meltdown of 2008, Iceland has, indeed, embarked on a major overhaul of its foundational text. As of July 2011, regular Icelanders have thus been invited to contribute ideas via the Internet and social media platforms such as Facebook, Twitter, YouTube, and Flickr, to supplement the work of a constitutional council, which regularly posts drafts on the Internet.[11] Whether these experiments are successful or not and whatever their flaws, they reflect the fact that the idea of collective intelligence has become mainstream and speak to its appeal, particularly in times of crisis.

In light of these evolutions, I now see my book as a product of its time, gathering a "knowledge" already present in popular and academic culture, albeit distributed over many sources and many individuals, and expressed in so many different forms as to lose clarity. It gathers, organizes, and synthesizes this common and distributed tacit knowledge and turns it into a coherent argument for democracy. My hope is that it also adds to this knowledge by putting forward a new argument in favor of democracy based on the correlation between inclusive decision making and cognitive diversity.

[11] See, e.g., http://www.theatlanticwire.com/global/2011/06/iceland-crowdsourcing-its-new-constitution/38713/ (accessed August 20, 2011).

Democratic Reason

CHAPTER ONE

The Maze and the Masses

DEMOCRACY IS GENERALLY HAILED, in the West at least,[1] as the only legitimate form of government. We (Westerners) only consider legitimate those regimes that are democratic or in the process of becoming so. Conversely, anything undemocratic raises our suspicion. In fact, democracy has such positive valence that some have argued that it is more than a descriptive term objectively referring to a certain type of regime or system of government; in this account, it is an evaluative term, by which its users commend certain institutions and societies (e.g., Skinner 1973). Even if we take democracy to simply denote a certain type of rule characterized by popular participation, it is a fact that this rule has, in the Western world, a privileged aura of legitimacy that competing rules are lacking. At the same time, however, there exists among contemporary democratic theorists, and even among the people themselves, a widely shared skepticism about the capacity of the people for self-rule.

The idea of democracy as a competent regime, in the sense of intelligent or even wise, is not intuitive. Ironically, indeed, democracy could easily be construed as the rule of "the idiots" (the Greek *hoi idiotai* stand for "the ordinary citizens"). In the philosophy of thought, political intelligence is generally a quality attributed to aristocracies, monarchies, and other elitist regimes—not democracies. If democracy is valuable, most people would argue, it is for reasons that have nothing to do with the intelligence, let alone the wisdom, of the masses. Indeed, these reasons have to override the notorious fact of the "folly of crowds."

Contra this widely shared intuition that democracy is a right that the people do not really possess the competence to exercise, this book consists in a sustained epistemic argument for democracy based on the idea of collective intelligence. I argue that democracy is a smart collective decision-making procedure that taps into the intelligence of the people as a group in ways that can even, under the right conditions, make it smarter than alternative regimes such as the rule of one or the rule of the

---

[1] In the following I consider my natural audience to be Western (or at least Westernized) political theorists and citizens in general. My excuse for such a narrow conception of the public worth addressing is that I would never have gotten started if I had to account for all the philosophical assumptions and principles underpinning the argument of this book.

few. This idea of collective intelligence offers, in theory, an attractive solution to the problem of the average citizen's ignorance and irrationality. If the many as a group can be smarter than any individual within them, then political scientists need not worry so much about the cognitive performance of the average voter and should focus instead on the emergent cognitive properties of the people as a group.

The argument, like many arguments in political theory, is of course not entirely new and has in fact a decent philosophical pedigree. It was first considered, with skepticism, by Aristotle (who himself borrowed it from the Sophists) and, as I will have the leisure to show at some length later on, has been running ever since in different guises as an underground current of political philosophy in the mainstream suspicion toward democracy. In contemporary democratic theory, as I will review in chapter 2, different versions and parts of the argument have been recently taken up by both political scientists and normative political philosophers seeking to use them for the justification of democracy.

This book offers an overview and assessment of the arguments that can be advanced in favor of what I call "democratic reason" or the collective intelligence of the people. I use these arguments, in connection to a literature on the wisdom of crowds, to support a strong epistemic case for democracy. I also develop the argument in a comparative manner, contrasting the epistemic benefits of democracy with those of nondemocratic decision procedures. Although I proceed essentially from an a priori and theoretical perspective, reasoning about democracy as an ideal type and a model, the argument that democracy is an epistemically superior form of decision making can be translated into an empirical claim that lends itself to falsification and can be supported by empirical evidence. The kind of empirical evidence that would tend to support the strong epistemic claim I put forward can be illustrated by Josiah Ober's recent study of Athenian democracy (e.g., Ober 2010 and 2012), which establishes that the superiority of classical Athens over rival city-states was due to its ability to process the distributed knowledge and information of its citizens better than less democratic regimes.[2] The institutions

---

[2] I became aware of Ober's manuscript *Democracy and Knowledge* when it was still unpublished but too late in the process of writing my own book to integrate its content properly. Ober later kindly accepted an invitation to write an essay for the edited volume *Collective Wisdom* that Jon Elster and I had been putting together since 2008 (now published with Cambridge University Press in 2012). His work provides, in my view, a compelling historical illustration, if not a systematic demonstration per se, of both Page's (2007) thesis that cognitive diversity matters to collective intelligence and my conjecture that under some reasonable assumptions and all things being equal otherwise, democracy is epistemically smarter than the rule of the few.

accounting for the epistemic superiority of Athenian democracy include, for example, the Council of 500 and the practice of ostracism.

1. The Maze and the Masses

The heart of the book is thus a defense of the "collective intelligence" hypothesis in favor of democracy. I argue that there are good reasons to think that for most political problems and under conditions conducive to proper deliberation and proper use of majority rule, a democratic decision procedure is likely to be a better decision procedure than any nondemocratic decision procedures, such as a council of experts or a benevolent dictator. I thus defend a strong version of the epistemic argument for democracy.[3] In my view, all things being equal otherwise, the rule of the many is at least as good as, and occasionally better than, the rule of the few at identifying the common good and providing solutions to collective problems. This is so, I will suggest, because including more people in the decision-making process naturally tends to increase what has been shown to be a key ingredient of collective intelligence in the contexts of both problem solving and prediction—namely, cognitive diversity.

I will explain at length what I mean by cognitive diversity later, but let me for now illustrate the epistemic argument for democracy presented here with a metaphor, inspired by Descartes's thought experiment in the Discourse on Method,[4] which, for all its limitations, should prove enlightening.

Imagine a maze in which a group of people happens to be lost. This maze has an exit (perhaps even several). Clues as to the way of finding the exit(s) are written on the walls. The clues are written in different languages and sometimes coded in pictograms and equations. The clues are dispersed all over the maze, sometimes inscribed very high on the wall, sometimes very close to the ground. Some are written in small fonts, some in very large fonts. The group itself is a typical sample of humanity. It is thus composed of very different people. Some people are nearsighted and some are farsighted. Some are good at mathematics and some are good at languages. Some are very bright and some are not bright at all. The members of the group care about each other and they have only one good: to get out of the

---

[3] The weak version of the epistemic argument for democracy holds that democracies are only better than random at making decisions.

[4] In the interest of full disclosure, I admit it is also inspired by a less highbrow source: a Canadian horror movie entitled Cube (1997), involving a group of people with different cognitive skills who learn (the hard way of course) that their only chance of survival in the "cube" is to collectively solve riddles, which requires cooperation and a pooling of their respective talents.

maze, preferably together. Every time they reach a fork in the maze,[5] they have to make a decision as to which direction the group should take. What kind of decision procedure should the group commit to at the beginning if their goal is to maximize their chances to get out of the maze?

Let us assume for now that the commitment is final and that the decision rule cannot be renegotiated at every fork in the road. At the beginning of the journey, the group faces the following set of choices. A first option is for the group to flip a coin at every fork and let chance decide. At every fork, the group has only a 50 percent chance of getting it right. Let us call this decision procedure "random."

A second option is to let one person make the decisions for the group at every fork. This person could claim the role because people believe she has a special connection with an all-knowing God, or because she can plausibly claim to be an expert in maze solving, or for whatever other reason. The group can let that person self-appoint or actively elect her. We can generally label the rule of such a person a dictatorship. Even if the dictator is initially elected, her power is not subjected to the classical accountability mechanisms of democratic representation, such as the challenge of repeated elections after the first (per Manin 1997).

The group can also choose to let a small council of people make decisions for the group at every fork. Even if the members of the council did not impose themselves by force or cunning but were chosen by the rest as smart and capable of making decisions on behalf of the group, we will label this council of experts an "oligarchy." Here again, the absence of a system of periodic elections and other accountability mechanisms makes it impossible to see this option as even remotely representative, let alone democratic.

A fourth option is for the group to choose at every fork to decide as a group, through deliberation followed by majority rule. If the group is too large, representatives (whether chosen by elections or a lottery organized at regular intervals, for example, at every new fork in the road) can make decisions on its behalf. We will label this option "democracy." When the whole group is directly involved in the decision, we are dealing with direct democracy. When only a subset of representatives is directly involved, we are dealing with indirect or representative democracy. For now, we will assume that representative democracy is not fundamentally different from direct democracy.

The choice is thus between a random procedure, a dictator, an oligarchy, and a direct or an indirect democracy. This book argues that, for the

---

[5] I assume for the sake of simplicity that choices are binary or can be reformulated as a long series of binary choices. I address some of the difficulties with that assumption (including the issue of agenda manipulation) in chapter 7.

purpose of getting out of the maze as a group, the democratic alternative, whether in a direct or indirect form, is not just better than random but better than the idealized dictator and the oligarchy. If there were only one fork and thus only one choice to make, letting a dictator decide on that one occasion might be as good an option as choosing as a group, since if the group were lucky, it would pick as a decision maker the one person who actually knew the answer on that particular occasion. As the problem poses a *long series* of choices to be made, however, it becomes very unlikely that the person elected as a dictator at the beginning of the journey, however smart and informed she might be, will know the right answer at every new fork in the road. It is more likely that someone else in the group will know the answer, or only the group as a whole will know the answer. It is smarter, therefore, to keep as many people as possible involved in the decision every time the group arrives at a new fork.

The comparison of democracy with the small council might seem trickier. If the group elects its brightest members at the beginning and asks them to make decisions on its behalf, it seems that it would have a better chance to get out of the maze than if it involves everyone, including the not-so-smart people. As we will see, however, group intelligence is only partly a function of individual intelligence. It is also a function of cognitive diversity—roughly, the existence of different ways of seeing the world—which is a group property. If the brightest individuals form a small group in which everyone happens to be nearsighted (because brightness is correlated with extensive reading, for example), they will not as a group have as much cognitive diversity as a larger group. Consequently they will miss the clues written in small fonts on the walls, which perhaps the not-so-bright people with good vision would be able to pick up.

A more inclusive decision-making process, or one that would renew the pool of decision makers through regular elections, would in that case produce a better decision. The advantage of democracy over a dictatorship and even a small group of experts is strengthened the longer and the more complicated the maze is. There is practically no chance that one single person could resolve a long series of difficult puzzles. There is a better chance that a small group of experts could, but the probability is still small. The way to maximize the probability of getting all the answers right and finding the way out of the maze is to include everyone, whether directly or indirectly.

Notice that assuming that the people elect the dictator or the group of oligarchs stacks the deck against democracy in the first place. If the people are generally competent at making political decisions, they should also be fairly competent at choosing their leaders, which means that an elected dictatorship or an elected oligarchy would be the most effective forms of these regimes. In practice, dictatorships and oligarchies are usually imposed from the top down, which rarely ensures that the individuals who end up as dictators or oligarchs are the best that they can be. They are more likely to be violent and power hungry than to be effective at pursuing the good of the country. What I aim to show is that democracy is epistemically superior to the rule of the few not just when the few are traditionally defined (as unelected and generally self-appointed oligarchs) but in fact even when the few are elected.

The metaphor of the maze has its limits, to which I will return in the conclusion. Assuming for now that I succeed in rendering plausible the claim that democracy does at least better than a random decision procedure and also better than alternative nondemocratic procedures, one may ask, first, What is the gain in terms of the normative justification for democracy? The answer is: a lot. At a minimum, to the extent that the goal of a justification for democracy is simply to establish the value of democracy as an instrumentally desirable regime, an argument showing that democracy has strong epistemic properties can only be welcome. At a maximum, the argument can reinforce, if not altogether establish, the legitimacy of democratic authority.

Second, if the goal of a justification happens to overlap with the task of legitimation of democracy as a normatively desirable regime—if, in other words, we do not draw a strict distinction between justification and legitimation—then the epistemic case for democracy may prove crucial to establishing the normative authority of democracy. David Estlund has opened this path by arguing that the normative concept of democratic authority includes an epistemic dimension (1997; 2007). Against consent theorists who believe that legitimacy is strictly a matter of the consent of the people and has nothing to do with the reasons that may or may not justify a regime, Estlund's "epistemic proceduralism" reintroduces instrumental considerations in the legitimation of democracy. In his view, democracy would have no normative authority, that is, no right to claim obedience to its decisions, if we did not assume that it met a minimal epistemic threshold, which he sets at "better than random." Estlund, however, is not really interested in providing the actual proof that democracy can be expected to meet that minimal requirement. If successful, the case presented in this book will provide such a proof.

Finally, to the extent that the book aims to prove an even stronger claim—namely that democracy epistemically outperforms nondemocratic rules—its contribution is to offer an argument for democracy (and, possibly, democratic authority) even to people who do not share the Western faith in the right of people to govern themselves when this right is entirely disconnected from the question of their capacity for it. One could imagine, for example, a person who subscribes to the idea that political authority must include an epistemic dimension but does not think that nondemocratic regimes are by definition disqualified from claims to legitimacy and normative authority. For such a person, an important question is the following: assuming that all regimes meet a minimal threshold of epistemic competence, what is it about democracy that makes it more normatively authoritative or legitimate than an equally competent oligarchy of the wise? There is a readily available argument for such a skeptical reader: democracy is simply a smarter regime than the rest.

While the contribution of this book should thus be clear, let me also emphasize its originality, which has to do with the connection drawn between inclusive decision making and the role of cognitive diversity in the emergence of democratic reason. All sorts of democrats have argued that a diversity of points of view, and even active and passionate dissent, are healthy for a democracy. From John Stuart Mill's defense of social gadflies to Cass Sunstein's celebration of dissent (Sunstein 2003), the case has been abundantly made that the existence of heterodox thinkers— people who think differently, including in violation of the group's most fundamental norms—spur the body politic away from complacent rehearsal of dead dogmas and toward more creative thinking.

This case for the cultivation of different viewpoints suggests tempering any form of government with liberal rights, including, crucially, freedom of expression. It does nothing, however, to support more-inclusive decision making as such. Although James Surowiecki (2004) has made the case for more-inclusive decision making on the basis of the statistical properties of large numbers and the Condorcet Jury Theorem, he says little about cognitive diversity per se. He also tends to focus only on the purely aggregative side of collective intelligence, dismissing its deliberative aspects as counterproductive and conducive to polarization. What I propose in this book is an argument that explicitly connects the epistemic properties of a liberal society and those of democratic decision procedure. I argue that in an open liberal society, it is simply more likely that a larger group of decision makers will be more cognitively diverse, and therefore smarter, than a smaller group. I thus attribute the epistemic superiority of democracy not only to the sheer number of decision makers, but also to the qualitative differences that, in liberal open conditions, this great number of decision makers is likely to bring with it. For all its simplicity, this insight is, I believe, new and important.

The epistemic argument for democracy based on the centrality of cognitive diversity that I defend in this book has at least three characteristics: it is maximal rather than minimal; it is probabilistic; it is theoretical a priori rather than empirical (and yet can be translated into an empirical and thus falsifiable claim).

On the minimal version of the epistemic argument for democracy, democracy is at least as good as, and occasionally better than, a random decision procedure at making decisions, although it can be inferior to rule by the wise few or the lone genius. On the maximal version of the epistemic argument for democracy, democracy is at least as good as, and occasionally better than, any alternative decision rule. Between those two extremes lie many intermediary positions. By choosing to defend the maximal version, I take a risk vis-a-vis other epistemic democrats, who simply compare the possible epistemic properties of some democratic procedures with those of other democratic procedures, such as deliberation versus majority rule, or compare democracy with a random procedure.

Note that I am not proposing a full-blown theory of democratic authority. I am interested here only in the epistemic side of a justification for democracy, which is in my view a necessary component of such a justification, although perhaps not a sufficient one. The epistemic argument for democracy that I defend is thus freestanding and conceptually independent from arguments for democracy that emphasize, for example, the fairness of the democratic decision process. Whether I succeed or fail in arguing for the maximal epistemic claim, democracy would still be an attractive form of government, as long as I (or others) succeed in establishing the minimal claim that it does at least as well and occasionally better than a random procedure. Above this threshold, classical pure proceduralist arguments can apply.[6] As long as democracy does better than random—that is, it performs minimally well—the fact that it embodies our ideals of equality and autonomy arguably would make it a worthy choice.

Another way to say this is to insist that the maximal epistemic argument for democracy presented here need not be the primary, let alone the only, argument for democracy. In the history of thought, democracy has generally been primarily justified on nonepistemic grounds, through appeal to notions of justice, equality, fairness, or consent. In earlier ages, an epistemic argument was doomed to failure in the face of deeply entrenched prejudices against the masses and the lack of an empirical record demonstrating how well democracies can do. This historical legacy has left us with a set of reasons for valuing democracy that are largely nonepistemic.

However bold it may seem, the epistemic argument defended here is—and this is its second characteristic—only probabilistic. The claim that democracy is overall the smartest method for making group decisions

---

[6] By "pure proceduralism" here, I mean the opposite of pure instrumentalism, namely a position that considers that the value of democracy lies in some intrinsic properties of its procedures, rather than in the quality of its outcomes.

does not exclude the possibility that some democratic decisions will be mistaken, nor does it exclude the possibility that a particular democracy will do worse than a particular oligarchy. Some of the time, the argument implicitly acknowledges, a democratic decision will prove inferior to one that would have been imposed by a dictator or a group of aristocrats or experts. Some of the time, a democracy will fail where a dictatorship or an oligarchy would have succeeded. On average, however, and in the long run, the claim is that democracy is a safer bet than a dictatorship or even an aristocracy.

The third characteristic of the epistemic claim presented in this book is that it is theoretical and a priori rather than empirical. I thus rely on models and theorems to support my case for democracy, rather than on case studies or empirical evidence. The probabilities I am thus considering in this book are theoretical or a priori probabilities, not frequencies (that is, empirically observed regularities). Where I do resort to empirical examples, they are meant only to support the plausibility of the theoretical claim, not establish it. This theoretical and a priori bent of the argument stems from the nature of the object itself: I am interested primarily in the *ideal* of democratic decision making. The reason is that while it may seem more important to focus on ways to improve the reality of democracy, this practical task presupposes first a clearer understanding of the ideal and why it is an ideal in the first place.

Although it is a priori, the epistemic claim put forward in this book can, I believe, be translated into an empirical and thus potentially falsifiable claim, provided the theoretical assumptions are shown to have a real-life equivalent. Wherever possible, I try to show that the assumptions behind various theorems (for example the Diversity Trumps Ability Theorem or the Condorcet Jury Theorem) plausibly translate to the real world. The theory thus allows me to venture a prediction: decisions about public affairs made by the many should be probabilistically (this time in the sense of "statistically") superior to decisions made by the few. While I do not myself build the empirical case to support this prediction, I mention existing supporting evidence whenever appropriate and trust that more could ultimately be found, whether in the form of qualitative or quantitative studies, case studies of specific democracies (as in Ober 2010), or large-N studies of groups of democracies. Of course, a lot hinges on what the empirical standard of "intelligent political decisions" is taken to be. Aside from the conceptual problems attached to the idea of such a procedure-independent standard of rightness, which I address in chapter 8, construing an empirical equivalent raises technical and methodological challenges that I cannot touch on here.

2. On the Meaning of Democracy

In this book I use the term "democracy" in a specific way: as, primarily, an inclusive collective decision procedure, that is, a procedure for collective decisions characterized by the fact that it is inclusive, more or less directly, of all the members of the group for whom decisions need to be made. Definitions of democracy vary, from a mere set of institutions—for example, at the minimum, majority rule (e.g., Hardin 1999)—to democracy as a way of life (e.g., Dewey (1927) 1954). This book endorses by contrast a thin definition of democracy as "rule of the many," using the term to refer to both a specific decision rule—whereby the many are the ultimate, if not direct, decision makers rather than one or a few persons—and, more broadly, the form of government characterized by this decision rule.[7]

An important aspect of my definition of democracy as an inclusive collective decision procedure is that it applies to representative democracy as well. The epistemic argument presented in this book is meant to be valid for representative democracy, at least on a democratic rather than elitist interpretation of representative democracy, such as the one proposed by Nadia Urbinati (2006).[8] In fact, in light of some of the risks of epistemic failures presented by direct democracy even where it is feasible, it is most likely that representative democracy is a more intelligent form of democratic regime than direct democracy per se, not because it is less democratic, but because it is less immediate, allowing people time for reflecting on and refining their judgment. Then again, I will only address in passing the diachronic dimension of democracy (its ability to reflect upon itself and learn over time), and so in this book the epistemic comparison between direct and representative forms of democracy is only sketched. Some of what the research presented here actually suggests is that while increasing popular input and participation in political decision making is generally a good thing from an epistemic point of view, this increased popular input does not require bypassing representation altogether—far from it.[9]

---

[7]Defining democracy principally as a decision rule rather than, say, a set of political institutions, a social ideal, and so on, is not standard. This definition, however, is faithful to the original meaning of democracy and one in fact used implicitly in the literature on deliberative democracy, particularly by epistemic democrats. See, for the latest example, Goodin (2008), who deals explicitly with deliberation as a decision procedure.

[8]In Urbinati's view, the representatives ought to be conceptualized as "advocates" of the people, defending their interests, expressing their concerns, and taking their judgments into account. They should not be seen, as in elitist theories of democracy, as an elected class of aristocrats whose judgment is entirely independent of that of the people, or dependent on it only to the extent that the representatives fear the voters' sanction at election time.

[9]I agree with Urbinati's view of representative democracy as an even "truer" democracy than direct democracy itself, although my conclusion is based on epistemic reasons not considered by Urbinati.

The definition of democracy as the rule of the many is meant to be contrasted with the rule of one and the rule of the few—two undemocratic forms of decision making that will serve as the comparison points throughout the book. The rule of the many is also distinct from the rule of all, in which unanimity would be required on all issues. I do not consider this form of collective decision making, as it is largely impractical.

A decision rule involving the many refers both to deliberation and to majority rule, which many theorists often pit against each other as corresponding to two different conceptions of democracy, one "deliberative" and the other "aggregative." In reality, democracy, whether direct or representative, always includes those two complementary procedures, which apply to two distinct contexts: problem solving in the case of deliberation, prediction in the case of judgment aggregation. Deliberation and majority rule thus feature distinct and complementary, rather than rival, epistemic properties. The epistemic properties of deliberation stem from the fact that it ideally allows for the identification or the construction of the best solution to a given problem or, in the terms of deliberative democrats, the triumph of the better argument. Where deliberation fails to produce a definite decision, it contributes to determining the alternative solutions between which a vote can then be taken as a way to predict which is likely to be the most effective. The epistemic properties of vote aggregation result from the law of large numbers and the right combination of individual accuracy and cognitive diversity, which deliberation can help set up.

My approach thus cuts across the traditional divide between deliberative and aggregative democracy in two different ways. First, I do not define democracy by reference to deliberation only or majority rule only, as if they were two competing democratic mechanisms. Second, contrary to many deliberative democrats and most aggregative democrats, I do not consider majority rule as a mere way to aggregate preferences or adjudicate fairly between competing interests. In my view, majority rule has its own distinct epistemic properties.

In comparing the three competing regimes—dictatorship, oligarchy, and democracy—I idealize them to assume that each rules "for the people," namely the entire community, as opposed to the interests of the rulers themselves or a subgroup within society. I do so in order to avoid complicating the comparative issue of epistemic competence (the kind of knowledge one has) with the problem of moral competence (the kind of intentions one has). I want to argue that, of the three regimes, democracy is the most epistemically efficient decision rule. The claim is meant to be valid with all things being equal otherwise—that is, controlling for potentially interfering factors like, most importantly, the problem of the moral integrity of the rulers. To make the comparison, all three regimes need to be assumed to share the same goal: that of pursuing the public interest as opposed to the interest of the rulers or some other partial interest. For some readers, this might make the epistemic argument defended here ultimately irrelevant, since surely the motivations of the rulers always play a role in reality. The analytical gain in conceptually separating the two problems, however, should be clear. Further, since the epistemic problem and the moral problem ask for different answers and institutional solutions, the more we can disentangle the issues, the more likely we are to fix them.

This comparison also treats all three idealized regimes as "pure" forms of government. In reality, these regimes generally include components borrowed from the other ideal types. Even the most confident of dictators surrounds himself with a group of advisors whose influence might mitigate the dictatorship with oligarchic principles. Conversely, real democracies include oligarchic or aristocratic dimensions along with genuinely democratic ones. Bernard Manin (1997) has argued that representative democracy is, historically speaking, not fully democratic but a "mixed" regime, because it includes democratic components (majority rule) and at least partly aristocratic ones (the representatives as an elected "elite"). This view of representative democracy as a mixed regime, however, is meant to be descriptive rather than normative. Even if this description is accurate from a historical point of view, it does not therefore stand in opposition to the normative ideal of representative democracy as essentially democratic and characterized by "advocacy" (see also Manin and Urbinati in Landemore 2008).

More importantly, I do not think that taking regimes as "pure forms" raises a problem for the validity of the epistemic argument for democracy. On the contrary, it makes it easier to understand what epistemic properties in existing mixed regimes can be attributed to their democratic, oligarchic, or dictatorial elements. It might be the case, in the end, that a mixed regime is the truly smartest form of collective decision making, although this is not a question I will try to answer. My focus is on pure forms of decision making.

Finally, let me point out that as this reflection on the nature of democracy applies to democracy as an ideal type, it is meant to transcend geographical as well as historical contingencies. Although I will occasionally rely on examples borrowed from the cases of the United States and France—the two countries I know best—such examples are meant to serve as springboards for a better understanding of the ideal of democracy, not so much the empirical realities of democracies around the world (which generally fall short of the ideal and thus make it very hard to think clearly about why we value democracy in the first place).

3. The Domain of Democratic Reason and the Circumstances of Politics

This book takes for granted the existence of a political domain of questions in contrast with a moral, scientific, or aesthetic one. I assume that the core of politics is the domain of questions where human beings deal with the risk and uncertainty of human life as a collective problem. This makes it by definition a domain where expertise is difficult to identify *ex ante*. In chapter 3, I will argue that the famous Athenian defense of an equal right of speech in the assembly—the reason why any citizen is allowed to speak on public matters in a way they would not be on more technical issues—can be interpreted in this light. In that sense, politics is unlike administration, precisely because, unlike for administrative tasks, when it comes to solving political problems, we cannot tell who the experts are. Administrative tasks are those tasks for which we can a priori determine whose knowledge and opinions matter. In politics, however, as in the maze, we do not have this luxury. Most of the time, because of the unpredictable and ever-changing nature of the problems that the community will have to deal with, the relevant knowledge, perspective, or information are simply unknown. In a democracy, we deal with this collective problem of risk and uncertainty as a group, rather than letting a king or an oligarchy solve the problems for us. Given the nature of politics, this is indeed the smartest solution.

Even though it is plausible that collective intelligence applies in morals, science, and the arts as well, the epistemic argument for democracy is meant to apply to politics only. I thus take for granted the classification by most societies of certain forms of knowledge as pertaining to different spheres. I specify this to make clear that the epistemic argument for democracy does not imply that democratic decision procedures should be used to decide any question, effectively turning every factual and value question into a political one and every political question into a democratic one. The epistemic argument for democracy need not entail radical implications and may simply support the claim that decision procedures involving the many are a smart way to make decisions wherever we already use such procedures. On that conservative reading, the implication of the book would simply be to stress the epistemic advantages of those democratic procedures as a worthy addition to their fairness.

Because there is no obvious reason why we use certain kinds of decision procedures in some cases and not in others, however, it would be surprising if the conclusions supported by the argument happened to be conservative, reinforcing the status quo rather than questioning it (unless we already live in the best of all possible worlds and our current practices are perfectly adaptive). We should be able to define the best boundaries of our practices and make more meaningful distinctions.

The epistemic argument for democracy thus can, and I think should, be interpreted in a more radical way, suggesting that we have epistemic reasons to expand the scope of questions decided by democratic means to, for example, a number of issues currently decided by experts. It may possibly even support the case against nondemocratic authorities outside of politics, such as that made by advocates of workplace democracy (e.g., Pateman 1970; Gould 1988; Greenberg 1986; McMahon 1994). A minima, and staying within the realm of politics, the argument would certainly suggest increasing popular input in questions as apparently technical as those related to nuclear energy, genetically modified organisms, pension systems, or electoral reform. Empirical evidence here tends to support this a priori conclusion. Experiments run in different countries putting regular citizens in charge of such questions show that average citizens can, under the right conditions, collectively produce smart contributions and proposals even on extremely technical debates. Political scientists and sociologists have thus documented in the last ten years the success of "hybrid forums" mixing experts and lay people, as well as that of the Danish "consensus conferences," also implemented in Japan and France as "citizens' conferences" (Callon, Lascoumes and Barthe 2001). Even more compelling evidence of popular wisdom can also be found in Citizens' Assemblies (Warren and Pearse 2008) and James Fishkin and Robert Luskin's (2005) deliberative polls—the latter arguably forming the "gold standard" of existing attempts at measuring what an informed public opinion would look like (Mansbridge 2010a). In all these experiments, experts admit being impressed by both the quality of the discussions and the nuanced conclusions reached by groups of self-professed amateurs.[10] The results of these experiments thus suggests that including more popular input would at worst not harm the quality of the decisions and at best enhance them. On that radical reading, the epistemic argument blurs the line between so-called technical and political questions, as well as between experts and amateurs.

In order to support the claim that regular citizens do at least as well as experts, one would of course need to be able to compare the political judgment of alleged political experts with that of the people. The potential trade-offs between considerations for procedural fairness and epistemic competence cannot but be decided (in part) on the basis of difficult

---

[10] This is not meant to imply that expert judgment is the standard of the right decision in those situations. Expert judgment here simply provides an external point of view that lends plausibility to the idea that laymen can contribute even to technical, difficult situations.

empirical questions of comparative competence. Until recently, the literature did not offer much in the way of a comparison, partly because democratic theory has not been traditionally interested in the concept of group intelligence and partly because we have yet to experience fully the limits of people's intelligence on a number of political issues. There is, however, an important literature on jury deliberations that can be used to support the democratic case, as well as a now-massive literature in deliberative democracy on the merits of group deliberation. Some of this literature will be considered in chapters 4 and 5. On the more pessimistic side of the existing research, one finds the "enlightenment preferences" literature, in which the researchers statistically simulate the views that regular citizens would hold if they were given the knowledge of a PhD and show that statistically enlightened citizens have views much closer to those of experts, seemingly proving the point that real voters are incompetent. Of course, this type of research is hardly empirical, but its conclusions are worth reviewing, which I shall proceed to do in chapter 7. The fact is that in most existing democracies, the skills of the citizens are presumed to be limited to choosing political candidates or answering referendum questions. Surely there is a vast range of other political questions for which the people as a whole might prove smarter than we currently think.

The competence of experts, on the other hand, has been the object of more interest and scrutiny. I will say more on this in chapter 8 of this book, as the attempt to define and quantify a good political judgment raises important questions for the position that I call "political cognitivism" (roughly, the view that there are right and wrong answers in politics and that these answers can be known, if only approximately). For now I will simply mention one of Philip Tetlock's conclusions regarding the characteristics of good political judges. When it comes to assessing a problem and making political predictions, Tetlock argues, political "experts" hardly do better than lay people and, on the purely predictive side, are in general outperformed by simple statistical regressions (Tetlock 2005). Other findings reported in chapter 6 of this book suggest that the aggregation of predictive schema in large groups of cognitively diverse people yields a fairly good approximation of complex statistical regression models. If, as the findings in this chapter suggests, what matters for predictive accuracy is the way experts think, and if the way the best experts think as individuals is comparable to the way some groups tend to think just because they are groups (i.e., cognitively diverse), then a single expert who thinks like a "fox" rather than a "hedgehog"—that is, in Tetlock's terminology, an eclectic thinker rather than one driven by a single big idea—is probably equivalent to a diverse enough group of people. In fact, diverse enough groups can outperform even a "fox," since the fox is outperformed by statistical regressions, and large groups' predictive powers can be uncannily close to those of a complex statistical regression.[11] In other words, if political foxes think like groups and groups can beat foxes, why not rely on them rather than on experts, including when those experts are foxes?

Another noteworthy fact in Tetlock's study is that the standards by which Tetlock ultimately assesses political judgment—empirical accuracy and logical rigor—are both subjected to the metastandard of a consensus of reasonable people. In his preface, Tetlock specifies that the standards for judging judgment are meant to "command assent across the spectrum of reasonable opinion" (Tetlock 2005: xvi). Tetlock's standard to assess the quality of a good political judgment is thus something like an intersubjective agreement among people of reasonable opinion, not the expertise of a single individual. People of reasonable opinion are not necessarily experts; they are anyone endowed with common sense and an ability to distinguish between good and bad judgment. It might take an expert like Tetlock to quantify good political judgment but not to recognize it.

Consider finally that what many experts do is give advice based on complex regressions. But even the best regressions designed by the best experts still need interpretation. So what is best? Let the expert, who represents but one view and one interpretation, decide alone on the meaning? Or have a larger and diverse public, with its multiple perspectives, decide as a group on the meaning of that regression?

Thus, in answer to the question "Where does the concept of democratic reason apply?" my own suspicion is that it applies everywhere citizens think they are capable of making decisions. They might actually be good even where they do not yet trust themselves for lack of an opportunity to exercise their judgment. In any case, it seems to me that the burden of proof should be on skeptics—especially those who insist on calling themselves democrats—who question the capacity of citizens.

This radical perspective does not mean that a division of cognitive labor is unnecessary. In complex industrialized societies, where time is scarce and delegation of authority a necessity, professional politicians and experts will always be needed. The recognition that a category of individuals labeled as "experts" can legitimately speak for the people is not necessarily incompatible with the view that ultimately the people know best and remain the underlying source of normative authority, not just on fairness grounds, but also on epistemic grounds. While the vagueness of the domain specification may frustrate some readers, consider that it

---

[11] See Scott Page's Free Lunch Theorem (2007), presented here in chapter 6.

would be self-defeating for an epistemic argument for democracy—one making the claim that the many generally know better—to suggest that one particular individual (even an "expert" like a political theorist or philosopher) is best situated to determine in advance where the many can or cannot be smart. The history of democracy is one of increasing enfranchisement, as women, nonwhites, the propertyless, and other second-class citizens became at long last trusted in their capacity to vote and participate responsibly in political life. If anything, history suggests that existing regimes, including democracies, have yet to explore the full potential of democratic reason.

This radical perspective is not either a denial that there exists such a thing as "democratic unreason." I devote two full chapters (chapters 5 and 7) to the conditions under which deliberation and majority rule can epistemically fail, and I consider several historical examples of democratic failures in the course of this book. In my view, examples of democratic failures are no worse than examples of tyrannical or oligarchic failures, although I cannot substantiate that claim here. Political scientists would be well inspired to write a black book of democracies that would allow a comparison with the black book of oligarchic regimes.[12] I suspect that in terms of the sheer number of their victims, democracies at worst match oligarchies.

I will consider in the conclusion the temporal dimension of democratic reason and suggest that democracies' capacity for learning and self-correcting over time is also a way to fight the risk of democratic unreason. We will see that the notion of "distributed intelligence" proves useful to conceptualize democratic reason as a process distributed over multiple generations and the institutions these generations have created. I do not spend much time on this aspect of the argument, however crucial I believe it to be, reserving it instead for future work.

4. Democratic Reason as Collective Intelligence of the People

The final contribution of this book is the concept of democratic reason itself. I use that expression as a label to capture the idea of the collective intelligence of the people. I further extend it to include the larger concept of "collective wisdom," when the dimensions of "learning over time" and "learning from one's mistakes" are taken into account (see also Landemore and Elster 2012).

---

[12] In that spirit, Michael Mann's (2005) study of ethnic cleansing performed by contemporary democracies only fulfills one part of the task, that of documenting what his title calls the "dark side of democracy."

Over the last several decades, cognitive science has shown that intelligence is a complex notion, encompassing diverse mental and social phenomena such as learning and understanding, reasoning and problem solving, perception and adaptation. This general definition has the merit of being valid across a range of cultures and populations.[13] Intelligence is thus distinct from "merely book learning, a narrow academic skill, or test-taking smarts," reflecting instead—"a broader and deeper capability for comprehending our surroundings—'catching on,' 'making sense' of things, or 'figuring out' what to do" (Gottfredson 1997: 13). The main theories of intelligence today (those of Gardner 1983; Sternberg 1985; and Salovey and Mayer [1990] 1998: 5) all emphasize the multidimensionality of intelligence, in contrast with narrower notions such as rationality.[14]

Collective intelligence is this concept of intelligence applied to groups as opposed to individuals. Although it can theoretically be a linear function of individual intelligence (the sum of the parts), collective intelligence is often conceptualized as an "emergent" property (more than the sum of the parts). In other words, collective intelligence is more than a function of individual citizens' intelligence and depends on properties that cannot be found in individuals themselves but only in the whole. Such a concept is often used to describe the behavior of groups of social animals such as ants or bees, which display a form of intelligence at the level of the group that is not found at the level of each distinct animal—also referred to as a "hive mind."

There are many reasons—philosophical, political, and methodological—why it may seem problematic to attribute a property like intelligence to a group, as opposed to an individual. Yet a growing literature in philosophy considers conditions under which groups are sufficiently integrated to produce outputs that we normally associate with rational agency—what I will call more generally intelligence (e.g., Rovane 1998; Pettit 2003; List and Pettit 2005a, b). Roughly, at a sufficient level of integration, it is pragmatically and explanatorily useful to describe the group's outputs in intentional terms, namely as the group's "beliefs," "judgments," "commitments," or "knowledge" (Dennett 1987). While a crowd in Leicester Square usually will not qualify as sufficiently integrated, the members of a team, an orchestra, or the citizens of a given country may qualify, provided that they share some values and some goals. The political reasons to look suspiciously at holistic explanations and a metaphysics of groups

---

[13]It is a better starting point in any case than Edwin Boring's (1923) definition of intelligence as "what intelligence tests test," which is circular and is falling out of fashion for that reason.

[14]Rationality as coherence of preferences and beliefs is at best one element of intelligence, and perhaps not even a necessary one (especially in the case of Salovey and Mayer's [(1990) 1998] notion of "emotional intelligence").

rather than individuals are understandable given the political implications often (unduly) derived from such approaches. More importantly, there are good scientific reasons to want to stick to the principle of methodological individualism, that is, the effort to explain all human actions in terms of individual rather than group agency.

Notice, however, that using the idea of collective intelligence to account for democratic epistemic superiority does not necessarily mean abandoning methodological individualism, as long as collective intelligence is an umbrella concept for a series of mechanisms that turn individual inputs into collective outputs. Further, the idea of collective intelligence does not imply that no intelligence is to be found at the level of individuals, but only that the emerging product has a different nature or quality from that of the individuals.

A second lesson from psychology and the cognitive science of the last few decades is that intelligence can be "distributed"—that is, stretched over, not divided among, "mind, body, activity, and culturally organized settings (which include other actors)" (Lave 1988: 1). Specifically intelligence can be stretched over what is called "cognitive artifacts," namely devices that help us accomplish complicated tasks. Language, inscriptional systems for representing language, maps, lists, and calculators are all examples of such cognitive artifacts. According to the exact definition, cognitive artifacts are "those artificial devices that maintain, display, or operate upon information in order to serve a representational function and that affect human cognitive performance" (Norman 1991: 17).[15] Distributed intelligence thus refers to an emergent phenomenon that can be traced not to individual minds but rather to the interaction between individual minds and between those minds and their environment.

Although the idea of distributed intelligence relies on a spatial metaphor of distribution across people and environments, it has a diachronic dimension, too. To the extent that (some) cognitive artifacts allow cognitive processes to be broken down into successive tasks, one can say that they distribute cognitive effort not just through space but also through time. For example, the principle of a list as memory enhancer divides the cognitive process of remembering into at least three chronological steps:

[15] A classic example of a cognitive artifact is the spatial arrangement of digits that help us perform a complex multiplication (Rumelhart et al. 1986; Wertsch 1998). One of the properties of cognitive artifacts is that they contain in themselves the knowledge required to solve a given problem. All the individual problem solver has to know is how to use the cognitive artifact, for example, a calculator. An important point to grasp about cognitive artifacts is that they "may enhance performance, but as a rule they do not do so by enhancing or amplifying individual abilities" (Norman 1991: 19). Individual abilities generally stay the same. What is expanded and enhanced are the "cognitive capabilities of the total system of human, task, and artifact" (ibid.).

the construction of the list, the mental action of remembering to consult the list, and finally the actual reading and interpretation of the items on the list (Norman 1991: 21).[16] Distributed intelligence thus usually comprises both a spatial and a temporal dimension.

At the individual level, the theory of "distributed cognition" puts into question the metaphor of the brain as a ruling and all-computing entity. The cognitive scientist Andy Clark (1998) argues that most of what we call "thinking" occurs as an interaction between (1) many different modules in the brain, (2) our actions, and (3) our direct environment. Even the activity of coming up with new "ideas" usually happens in a piecemeal way, as the result of constantly moving back and forth between the reading of old notes and the production of fragmentary thoughts that connect those old thoughts with new ones, leading to more marks on the paper, which in turn suggest new trains of thought. Even though the brain has an essential function, it can accomplish very little by itself.[17]

A famous study by Edwin Hutchins (1995) showed how the study of cognitive processes must go "beyond the skin" and consider both how individuals and groups of individuals use historically evolved artifacts in carrying out their activities. Studying the way in which a military ship's complicated maneuvers are performed, Hutchins found out that the meaningful cognitive unit was not any single individual member of the crew but the navigation team as a whole. Hutchins observed that the computation involved in conducting a large ship takes place not in the head of any individual, but in a coordination of the minds of different individuals with navigational artifacts, such as landmarks, maps, phone circuits, and organizational roles.[18]

Hutchins's case of a large ship is particularly suggestive from a political point of view, because it challenges the antidemocratic implications usually derived from the comparison between leading a state and steering a ship. As we know, Plato used that analogy to antidemocratic ends, to counter the claim of the many to rule. In his view, only the philosophers were good helmsmen. The power-hungry masses of the ignorant, he assumed, were bound to wreck the ship. Plato's use of the analogy

[16]Hutchins (1995; more on whom below) calls this distribution of cognitive efforts across time "precomputation."

[17]This is why, according to Clark, most of the attempts to create artificial intelligence by focusing on a reproduction of only the brain have all failed.

[18]Notice that in this example, the cognitive tasks performed by individuals themselves are relatively simple. In the same way that multiplying two large numbers does not demand much cognitive effort on the part of the individual equipped with the right cognitive artifacts, the cognition of individuals on the ship often involves little more than reading numbers or drawing lines. The agent performing the symbol processing necessary for navigation is the system as a whole (a collection of individuals acting with cognitive artifacts), not single individuals.

implies that the kind of intelligence (or, in his vocabulary, "wisdom") required for conducting both ship and state can be located only in one or a few individuals. Hutchins's description of what actually takes place aboard modern ships suggests that Plato might be wrong on both points. If intelligence is "distributed," then having the many participate in ruling might not be such a bad idea for the purposes of steering both a ship and a state.[19] Thus, even if no citizen, not even the president, can truly be said to possess all the political knowledge necessary to make good political decisions, the ship *Democracy* nonetheless floats. I will also argue that it goes overall in the right direction (even if the right direction simply means "away from trouble").

While individual intelligence is thus not simply located in the brain but distributed across our actions and immediate environment as well, the collective intelligence at work in certain human organizations is also distributed, in the sense that it is not located in one central entity but stretched over many individuals and the cognitive artifacts that are parts of their environment.

Such a notion of democratic reason as distributed collective intelligence of the people is relevant for the epistemic argument for democracy in at least three ways. First, the idea of collective distributed intelligence is particularly useful to describe democracy as a system channeling the intelligence of the many and turning it into smart outputs. Democratic reason denotes a certain kind of distributed collective intelligence specific to politics, in contrast to the undemocratic kind of collective distributed intelligence observed by Hutchins on a military ship, that of the market, or that observable in a group of social animals such as bees or ants.

The concept of collective distributed intelligence also helps explain how the individual citizen cognitively unburdens him- or herself by letting others, as well as the environment, process parts of the social calculus. From that point of view, the idea of democratic reason as collective distributed intelligence offers an answer to the apparent paradox of the

---

[19] This is not to say that a military ship is an example of "democratic" decision making. The point of the example is only to suggest that even in as hierarchical an organization as a military ship, the knowledge required to make it function is distributed across more people than just the leaders. To the extent that a more democratic organization is capable of tapping into that distributed knowledge more efficiently, we may derive democratic implications from the example. This might not convince the army to change its authoritarian ways, but the argument seems to have somewhat pervaded the world of firms, where theorists of organization have observed a flattening of hierarchical structures—to minimize the degrees that information has to travel between Bill Gates and the rank and file, say—and a greater amount of delegation to the lower levels, precisely for those "cognitive reasons" (e.g., Mandeville 1996). In general, the idea of collective intelligence applied to politics does not preclude the cognitive division of labor that we observe on a boat or in representative democracies (where elected representatives specialize and arguably know more than the represented).

right of the people to rule themselves and the simultaneous belief (documented in chapter 2) that they lack the cognitive competence for it.

Finally, combined with the concept of cognitive artifacts that contain the wisdom of the past, the idea of a collective intelligence distributed not only through space (over people and artifacts) but over time as well introduces a temporal dimension into the concept of democratic reason. Democracies can learn, particularly from their own mistakes, how to immunize themselves against the worst forms of cognitive failures and how to embody in durable institutions the lessons learned from such past failures and mistakes. I will go back to that idea in the conclusion of this book.

I will occasionally use the concepts of reason, intelligence, and wisdom interchangeably, even though there are distinctions between them. Reason is the umbrella concept, referring to the general faculty through which human beings make choices and decisions that can be reasonable, intelligent, or wise. Reason denotes several concepts, such as the moral notion of autonomy, the cognitive notion of intelligence, as well as the moral/ experiential notion of wisdom. Intelligence, by contrast, denotes only the cognitive side of reason and does not always overlap with wisdom.

I will occasionally use the term rationality as a proxy for intelligence, but only where the intelligence at stake is logical or analytical. Rationality denotes a computational ability to identify the relevant means to achieve given ends. Intelligence, by contrast, has a wider definitional spectrum, denoting a property of the mind that encompasses many mental abilities besides the ability to think logically and abstractly. The term "intelligence" also has adaptive and creative connotations that are lacking in the common use of the term rationality. Finally, intelligence applies to emotions as well as rational faculties, allowing for a more flexible and richer understanding of the very notion of democratic reason.

Finally, I will use wisdom essentially to refer to the time-tested elements of collective decisions. Wisdom is a broad concept that I do not fully explore in this book, as it would entail taking a *longue durée* type of approach to democratic decision making, whereas I focus here mostly on the synchronic epistemic properties of collective decisions, with the exception of the role played by the deliberative delay introduced by representative institutions.

The choice of the expression "democratic reason" to refer to the collective intelligence of the people is inspired in part by a desire to contrast this notion with the Rawlsian idea of "public reason" (Rawls 1993). Rawls's idea of public reason defines a norm of liberal justification that is paradigmatically expressed by a restricted number of political agents (such as representatives, candidates to public offices, or justices of the Supreme Court). As a result, public reason in Rawls often seems limited to the reason of an idealized public, in contrast with the type of intelligence that I argue can emerge from the aggregated views of the entire citizenry, including very often some of the unreasonable individuals in it. The Rawlsian concept of the "reasonable" itself is construed so as to constrain the content of the democratic will a priori and filter out a number of antisocial, selfish, or illiberal preferences.[20]

It is probably true that liberal constraints should limit the democratic will and filter out certain preferences. The appeal of the standard of public reason thus lies in some of its content, such as a scheme of basic rights and liberties, and general lines of inquiry about shared commonsense methods (Rawls 1993: 223). Nonetheless, I do not want to internalize those constraints too early into the epistemic approach to democracy. In contrast to public reason, "democratic reason" is supposed to be the end product of a successful democratic process (particularly in its deliberative dimension, which can transform selfish, antisocial, and uninformed preferences into publicly minded and informed ones), rather than a filter for it. I will thus use the term to refer inclusively to the intelligence of the public at large, not of some subset of reasonable people within it. If voices and views must be ignored at any point, I trust that this will be a conclusion reached at the end, not the beginning of a study of collective intelligence in politics. Such conclusions might even be part of a theory of liberalism that needs to supplement an account of democratic authority but cannot be generated within it.

By the term "democratic reason," I thus deliberately distance myself from that liberal, Rawlsian tradition in order to remain strictly focused on the epistemic dimension of collective political judgment and the connection between epistemic properties and truly democratic—that is, all-inclusive (among other characteristics)—decision making. The methodological choice to stay out of a Rawlsian framework of public reason also accounts for why I undertake a comparison between the epistemic properties of the rule of the few and the rule of the many. Unlike Rawls and epistemic democrats in the Rawlsian tradition, I think that the epistemic challenge to democracy presented by the rule of a wise few or the rule of experts needs to be taken seriously.

5. Overview of the Book

The book includes eight chapters and a conclusion. Following this introductory chapter, chapter 2 illustrates what I take to be the deeply

---

[20]Rawls's conception of "public reason," which has been extremely influential on deliberative democrats, also permeates David Estlund's theory of "democratic authority," to the point where the important stress Estlund places on the epistemic dimension of political authority is, in my view, somewhat undercut by the role assigned to the liberal standard of reasonableness defining what people can legitimately consent to (the general "acceptability requirement"; see Estlund 2008: chap. 3).

entrenched prejudice of political philosophers, including some democratic theorists, against "the rule of the dumb many." This chapter offers a critical literature survey showing how most traditional approaches to democracy either deny or circumvent the question of the people's competence to rule, with the exception of a tiny but growing literature on "epistemic democracy," of which this book is meant to be a part.

Chapters 3, 4, 5, 6, and 7 provide together the positive argument in favor of the strong epistemic properties of democracy. Chapter 3 is a historical detour useful to appreciate, through a selected sample of authors and works, the range of arguments developed throughout history in favor of the epistemic properties of democracy. Chapter 3 thus contrasts two types of authors, the "talkers" and the "counters," who respectively credit democratic deliberation or some version of judgment aggregation for the epistemic performance of democratic regimes. The chapter traces the history of such arguments from the foundational myth of Protagoras, in which every human being is assumed to be endowed by the gods with a spark of political wisdom, to Condorcet and his famous Jury Theorem, to John Dewey's idea of democracy as a social quest for truth and Hayek's theory of "distributed knowledge."

Chapters 4 and 5, on the one hand, and chapters 6 and 7, on the other, are devoted to one of the two mechanisms identified in the preceding historical chapter: respectively, inclusive democratic deliberation and majority rule. Both sets of chapters make use of Lu Hong and Scott Page's (2001, 2004, 2009, 2012; Page 2007) work on the role of cognitive diversity in the emergence of group intelligence and apply it to the specific case of democratic decision making, seen as a combination of deliberation and aggregation of judgments through majority rule.

Chapter 4 stresses that deliberation is best designed for problem solving, in which cognitive diversity turns out to be more crucial than individual ability. Because deliberation allows the weeding out of bad information and arguments from the good, it does not matter too much if the individuals taking part in the deliberation do not have maximal cognitive abilities. It matters more that the group is characterized by a high degree of cognitive diversity. If the group can be so characterized, the case can be made that including more people, which generally means including more cognitive diversity, will make the group smarter. The argument further translates into a defense of descriptive representation and the selection of representatives through random lotteries rather than election.

Chapter 5 considers several objections, some empirical, some theoretical, to the claim that deliberation has positive epistemic properties. This chapter, in part the product of a joint work with the evolutionary psychologist Hugo Mercier (Landemore and Mercier 2010 and Mercier and Landemore 2012), uses a theory recently developed in evolutionary psychology—the argumentative theory of reasoning—to refute at a theoretical level two classical objections to deliberation. The first objection is that, far from leading to any individual or collective epistemic improvements, deliberation with others does not do much to change minds. The other objection is the so-called law of group polarization (Sunstein 2003) according to which deliberating groups of like-minded people will systematically polarize. The chapter argues that where the normal conditions of reasoning are satisfied, dialogical deliberation of the kind favored by most deliberative democrats is likely to have the predicted transformative epistemic properties.

Chapter 6 argues that majority rule is a useful complement of inclusive deliberation, not just because majority rule is more efficient timewise, but because it has distinct epistemic properties of its own. The chapter also stresses that majority rule is best designed for collective prediction— that is, the identification of the best options out of those selected during the deliberative phase. Of all the competing alternatives (rule of one or rule of the few), majority rule maximizes the chances of predicting the right answer among the proposed options. The chapter considers several accounts of the epistemic properties of majority rule, including the Condorcet Jury Theorem, the Miracle of Aggregation, and a more fine-grained model based on cognitive diversity borrowed from Hong and Page.

Chapter 7 addresses various objections to the epistemic properties of judgment aggregation through majority rule, including those based on impossibility theorems and the doctrinal paradox. The chapter also addresses the problem of "informational free riding"—that is, the problem of voter's rational ignorance in the context of large elections where their vote is unlikely to make any difference. Finally, the chapter refutes the claim that because voters are "rationally irrational" (Caplan 2007) and systematically biased, no good can come out of their aggregated views.

Chapter 8 offers a defense of one of the main and most controversial assumptions on which the epistemic case for democracy relies, namely that there exists a procedure-independent standard of correctness in political matters and that this standard can be approximated by human decision making. I call this assumption "political cognitivism." This chapter deals with various forms of political cognitivism and the so-called fact/value dichotomy. It also answers the antiauthoritarian objection that assuming a common good or a general interest requires an authoritarian and dangerous notion of truth.

By way of a conclusion, I return at the end of the book to the metaphor of the maze and the masses introduced in the first chapter and address a few concerns about the possibility of democratic "unreason." Introducing the dimension of time and reflection over time, I suggest, first, that democracies can learn from their mistakes and, second, that certain democratic institutions and norms serve as cognitive artifacts that help the people control for or correct their potential cognitive failures. Those cognitive artifacts at the level of society include institutions (such as a democratic constitution) and norms (such as a societywide commitment to fundamental rights) that embody the collective intelligence of the people distributed across both space and time. Democratic reason thus includes the wisdom of the past "many" crystallized into social cognitive artifacts that help reduce democratic unreason. Because of the synchronic and diachronic collective intelligence tapped by democratic institutions, democracy, I conclude, is a gamble worth taking.

CHAPTER TWO

Democracy as the Rule of the Dumb Many?

> "A democracy is nothing more than mob rule, where fifty-one percent of the people may take away the rights of the other forty-nine."
>
> —Thomas Jefferson

> "What constitutes the state is a matter of trained intelligence, not a matter of 'the people.' "
>
> —Friedrich Hegel

> "Democracy substitutes election by the incompetent many for appointment by the corrupt few."
>
> —George Bernard Shaw

> "The best argument against democracy is a five-minute conversation with the average voter."
>
> —Winston Churchill

THE IDEA OF DEMOCRACY as the only legitimate regime is recent. While the idea slowly took root in the last hundred years, it has become fully accepted only since World War II.[1] There exists, by contrast, a long-standing prejudice against democracy, not only among political theorists and philosophers but, more surprisingly, among the people themselves, based on a general suspicion of people's capacity for self-rule. As Robert Dahl puts it, "the assumption that people in general—ordinary people—are adequately qualified to govern themselves is, on the face of it, . . . an extravagant claim," which explains why it has been rejected since the very birth of democracy over two thousand years ago (Dahl 1989: 97).

Much of Western philosophy thus easily reads as an attack on the incompetent multitude. Plato flatly despised the masses and thought politics better left in the hands of a few philosophically trained experts. Aristotle, although he benevolently reports the sophistic doctrine of the wisdom of the multitude and sometimes seems on the brink of endorsing it, ultimately did not really trust that the many could be as a whole wiser

---

[1] See Dunn (2005) for a general history of the word "democracy" and Rosanvallon (1993), for the history in France more specifically.

than the best few. Two thousand years later, as democracy was historically and conceptually on the rise, coming to refer no longer to an archaic
and potentially dangerous regime but to a modern egalitarian society and
a legitimate form of government (Rosanvallon 1993),[2] Tocqueville still
could not help but contrast the superior minds at the head of aristocratic
regimes with the general mediocrity of democratic rulers and developed
one of the most powerful arguments against democracy, the argument
of the tyranny of the majority.[3] Nietzsche, reprising Plato's hatred of the
masses, scorned democracy for the kind of humanity it thrived amongst
and in turn bred.[4] In the history of thought, philosophers and political
theorists have thus spared neither wit nor energy in defiling democracy
as the rule of the incompetent and the ignoramuses, contrasting it with
the supposedly superior wisdom of philosopher-kings, professional politicians, experts, or any other small number of real "knowers." This tension between democracy and philosophy, it must be said, goes back to a
fundamental democratic sin: Socrates's death. If Western philosophy is a
series of footnotes to Plato, it is no wonder that the general tone of these
footnotes has been one of distrust toward democracy.[5]

[2]In France, for example, the turning point for the shift in the evaluative meaning of
"democracy" can be traced back to the years immediately following the French Revolution.
Pierre Rosanvallon thus observes that the word was mostly absent from political discourse
until then and, when mentioned, essentially was used in reference to the archaic rule of
Athens and Sparta; "in 1794," however, "the word 'democracy' was heard as both a rallying
cry and a threat, referring equally to the active power of the people and to the pathologi-
cal forms of its outbursts." Since then, the word has slowly gained its letters of nobility,
becoming less associated with an archaic form of direct rule by the people and more with
the modern egalitarian society and the representative form of government. According to
Rosanvallon, the sociological connotation was acquired in the years between 1814—the
date of the restoration of the Bourbon monarchy—and 1835, with the publication of the
first volume of Tocqueville's *Democracy in America*. The semantic shift to a more political
meaning—democracy as representative government—took place after 1848 with the advent
of universal suffrage. From then on, Rosanvallon suggests, "the term 'democracy' quickly
became dominant, designating both a regime and a form of society" (Rosanvallon 1993:
27, Rosanvallon's translation).
[3]It is true, though, that one also finds in Tocqueville a praise of the American people as
on average more educated than the French. While the common people are typically better
educated than in France, he remarks, the elites are comparatively less brilliant, hence the
general impression that America's democracy is the rule of the unenlightened.
[4]Nietzsche thus sees Europe's democratization of Europe as leading toward "the leveling
and mediocritization of man" and, consequently, the enslavement of the democratic man
to a higher type of beings, which Nietzsche refers to as "tyrants—taking that word in every
sense, including the most spiritual" (Nietzsche 1989: 176–77).
[5]Suspicion about the possibility of something like democratic intelligence, however,
is not just a philosophical prejudice. The belief in the "folly of crowds" is, ironically, part
of popular wisdom as well. This belief derives from a general lack of confidence in group
thinking, particularly majority thinking. This lack of confidence is rendered plausible by
well-known popular blunders, a catalogue of which the conservative journalist Charles

Again this general trend, chapter 3 will try to identify what in Western philosophy appears as a democratic undercurrent and will provide a survey of the most salient proto-epistemic arguments for democracy. The point of this chapter, by contrast, is to document the reasons that explain the remaining skepticism, in contemporary political sciences, toward democracy.

While the resistance of past philosophers to the ideal of popular self-rule should perhaps not come as a surprise, the ambivalence in the discourse of contemporary theorists is more puzzling. This ambivalence characterizes in particular contemporary political science, torn as it is between normative claims about the legitimacy of popular rule, on the one hand, and the antidemocratic conclusions induced by dispiriting empirical findings about citizens' civic competence and no less pessimistic theories of citizens' irrationality or democracy's meaninglessness, on the other. With the exception of a small group of "epistemic democrats," the question of the cognitive competence of average citizens and the related question of the performance of democratic institutions either raises profound skepticism or is avoided altogether in contemporary democratic theory, both positive and normative. As a result, many theories and justifications of democracy tend to be competence insensitive, either denying that citizens' political incompetence is a problem or circumventing what they do see as a problem through an "elitist" definition of democracy as rule by the elected enlightened.

1. The Antidemocratic Prejudice in Contemporary Democratic Theory

In 1973, Moses Finley pointed out that while the vast majority of contemporary intellectuals claimed a democratic allegiance, most also agreed that none of the principles in the name of which democracy was justified worked in practice (as illustrated by the recent fate of democracies in Europe)—and that for the sake of democracy's survival itself, it was wiser to disregard these principles (Finley 1973: 8–9). In describing the state of democratic theory at the time, Finley targeted the then-dominant "elitist" theory of democracy, of which Seymour Lipset was, in Finley's eyes,

---

Mackay provided for example in his famous *Extraordinary Popular Delusions and the Madness of Crowds* ([1841] 1995). Among many popular follies, Mackay recounts financial bubbles, religious crusades, witch hunts, the fashion of poisoning, and admiration for Robin Hood. In the eyes of the contemporary many, one might argue that the crucial sins of democracy are, besides Socrates's killing, its tolerance of slavery both in ancient and modern times, and, arguably, its responsibility for bringing Hitler to power.

the leading figure. Lipset himself was following Joseph Schumpeter, for whom democracy is merely a method ensuring a competitive selection of leaders by the people, not the rule of the people themselves (Schumpeter 1942: 256). According to Schumpeter, "if results that prove in the long run satisfactory to the people at large are made the test of government for the people, then government by the people . . . would often fail to meet it" (1942: 256). In Schumpeter's view, the people are not the best protectors of their own interests; hence arises the necessity to delegate power to an elite group of better decision makers.

What Finley diagnosed, referring to the contemporary state of democratic theory, as a "curious, paradoxical situation" (Finley 1973: 9), some political scientists at the time saw as congenial to democratic politics. In a nice contemporaneous formulation of this point, Meyer argues: "Despite the various philosophical, logical, and moral claims that have made 'rule by the people' a powerful, if not unambiguous, modern idea, democratic politics has never satisfactorily coped with the basically empirical claim that 'the people,' leveled and free of social restraints, do not really know what is best for them" (Meyer 1974: 197). For Meyer, democratic politics is characterized by the following belief: not only are the people incapable of protecting their own interests; they can barely identify them. Operating with such an assumption of cluelessness on the part of citizens, democratic theory unsurprisingly ends up with a paradox, displaying on the one hand "an evocative and rhetorical acceptance of 'the people' in a variety of ideological forms as the source of authority and right" and yet at the same time contending that "the wants and desires of the common man are inextricably out of line with what he really needs and ought to have" (Meyer 1974: 202). The tension between the lip service officially paid to democratic rule and the general suspicion toward the competence of the people is illustrated, in Meyer's view, by the case of the Marxist, who "fears false consciousness" on the part of the people; the liberal, who "bemoans the inarticulateness of public opinion and recoils from the vision of the politics of mass society"; and the Third World nationalist who distrusts the people because of the remnants of colonialism and parochialism, which he believes impair their judgments (Meyer 1974: 202). Meyer's conclusion strikes an especially pessimistic note, denying that the theorists of his time actually demonstrate the necessary "willingness to put the future of society in the hands of any and all of its members which was such a courageous commitment of those through history who have been sincerely concerned with the implications of the democratic faith" (Meyer 1974: 202). Meyer concludes that in his era the democratic commitment is either insincere or unsupported.

Finley's and Meyer's converging assessments of the state of democratic theory thirty years ago is no less accurate today. The relevant oppositions in recent democratic theory may no longer be exactly that of Marxists, Third World nationalists, and liberals—but rather the oppositions between aggregative and deliberative democrats, to a lesser degree between proceduralist and constitutional (or substantive) democrats, or even between nationalist and cosmopolitan democrats (Shapiro 2003b).[6] The diagnosis of a paradox, however, remains accurate. Democratic theory has evolved in a more participatory and deliberative direction, involving, ideally at least, the public as an active political agent. Yet this normative commitment is generally disconnected from a belief in people's competence.[7]

It may be that there is, ultimately, no necessary tension between a defense of people's right to self-rule and a simultaneous belief in their incompetence for it, if one believes, in particular, that issues of political legitimacy can and should be entirely disconnected from instrumental considerations of the quality of policy making. Even bracketing the issue of legitimacy, however, and restricting our concern to the instrumental justification of democracy, there should be something profoundly disquieting for democracy's supporters in the widespread belief that people can't rule themselves. Let us now review the reasons, both empirical and theoretical, supporting this belief.

2. What's Wrong with the People?

The traditional aristocratic prejudice against democracy stemmed from the belief that the people were neither wise nor virtuous enough to rule. The contemporary version of the same belief is that the people are irrational, uninformed, and apathetic. As a consequence, just like former aristocrats, most theorists of democracy conclude that democracy cannot be

[6] According to Shapiro's account of "the state of democratic theory" (2003b), the opposition between aggregative and deliberative theories of democracy forms the most important watermark.

[7] Doubts about the competence of the people for self-rule is not just a belief shared by political theorists and social scientists in general. They occasionally characterize the mood of the public itself and when they do, they seem to undermine people's confidence in the value of democracy. In France, for example, a recent study shows that 41 percent of the respondents agree with the following statement: "What the country needs is a strong man who does not care about the Parliament or elections." The same study shows that 55 percent agree with the view that "It should be experts, and not the government, who decide what is best for the country" (Grunberg 2002: 121, my translation). Commenting on these figures, Philippe Breton wonders whether there is still a majority of democrats in France (Breton 2006: 18). Breton concludes that the people themselves doubt their capacity for self-rule. They are ready to delegate their right to rule to allegedly more competent strong men and/or experts.

called an intelligent regime, since neither the individual citizens nor their collective decisions are smart.

Theoretical and empirical arguments, within both the positive and the normative side of contemporary political sciences, routinely support the first claim about the individually "incompetent citizen."[8] Rational choice theory predicts that rational agents would not vote, while public opinion research shows voters to be ill informed and politically apathetic. Social choice theory supports the second claim, demonstrating a priori why collective popular decisions cannot make sense.[9] All the while, normative theories of democracy—not only Schumpeterian, but also participatory and deliberative theories—either solve the problem in a nondemocratic way or ignore it altogether.

2.1 Positive Approaches

2.1.1 FIRST REPROACH: CITIZENS ARE IRRATIONAL

The classical denunciation of citizens' irrationality rests on the observation that voters turn out to vote despite the causal inefficacy of an individual vote in large elections. The idea that a vote contributes too marginally to the outcome to provide an individual incentive for voting was first formalized in Downs's seminal work, *An Economic Theory of Democracy* (1957), which phrased it as "the paradox of voting." If people were rational, so the argument goes, they would not take the time to vote at all. Since we empirically observe that they do vote, we end up with a "paradox."

In many ways, the paradox of voting is a theoretical puzzle created by a strictly instrumentalist approach to voters' behavior, according to which it makes sense to vote only if you are likely to be pivotal to the outcome of the election. Noninstrumental accounts, however, have been proposed that emphasize citizens' sense of civic duty (Barry [1965] 1990) or their desire to express their political views (Brennan and Lomasky 1993; Brennan and Hamlin 1998). Even on an instrumentalist account, it has been argued that there is nothing paradoxical about voters' decision to vote. Richard Tuck (2008) thus argues that a vote is efficacious to the extent that it is an element of a set of efficacious votes. Gerry Mackie (2007, 2012, and 2008) proposes in his "contributory theory of voting" that voting has two values, one strictly instrumental—my vote causes my

---

[8] The term "incompetent citizen" is from Delli Carpini and Keeter, "Political Knowledge of the U.S. Public," p. 2, cited in Popkin 1994: 42.

[9] True, these pessimistic claims are made in terms of (lack of) rationality and low levels of information or political engagement—not intelligence. I will deal later on with the meaning of political and social intelligence, at both the individual and collective level. Suffice to say for the moment that when rationality, information, and engagement are denied, intelligence is unlikely to be asserted.

team to win when it is pivotal—and the other instrumental in a larger sense. Mackie calls "mandate value" this other instrumental function of voting, which consists in causing the favored team to win by a certain amount or lose by a certain amount. The mandate value of voting is what ensures that it makes sense for the individual citizen to vote in large elections even if she's unlikely to be pivotal. I say more on these new answers to the paradox of voting, which also work as answers to the problem known as "informational free riding," in section 2 of chapter 7.

The economist Bryan Caplan, for his part, deplores irrationality not in the fact that citizens vote but in the way they vote. According to Caplan (2007), voters have preferences based on beliefs and maximize the ideological pleasure of feel-good beliefs. Taking into account the lessons of psychology and behavioral economics, Caplan argues that citizens' judgment is afflicted by systematic cognitive biases such as "the antimarket bias" or the inability to understand that private greed can benefit the public good.[10] As a result of such biases, citizens cannot be expected to want the means to their own preferred ends. Thus, to get the people what they want, the last thing one should do is ask them for their opinion. In effect, Caplan argues, we should have much less democratic input, particularly on economic issues. In other words, we should let economists rule, or better still, the market. While the book presents itself as bold and provocative—its cover displays a flock of scarily identical sheep, suggesting the conformism of average citizens—its message is ultimately classically conservative. In the wake of a long antidemocratic tradition that runs from Plato through Schumpeter, Caplan reiterates the idea that some people (economists) know better than the people themselves. If the goal is to satisfy the people's interests, the subtext reads, democracy should give way to the rule of these knowers.

2.1.2 SECOND REPROACH: CITIZENS ARE APATHETIC

Another cause of concern seems to be the level of apathy observed among democratic electorates, where apathy refers to the lack of interest and engagement in public life on the part of large numbers of citizens. In the United States, participation in national elections from 1930 to 2004 generally oscillates between 49 percent and 63 percent of the voting-age population.[11] Even during the years of highest turnout (1950 to 1964,

---

[10] Other systematic cognitive biases Caplan denounces are what he calls the antiforeigner bias (a distrust of foreigners that leads to overly protectionist policies), the underestimation of the benefits of conserving labor, and the pessimistic bias (the wrong belief that the economy goes from bad to worse).

[11] Data obtained from "Participation in Elections for President and U.S. Representatives, 1930–2004," © 2000–2006 Pearson Education, publishing as Infoplease, accessed 2 December 2006, http://www.infoplease.com/ipa/A0763629.html.

where participation was consistently slightly above 60 percent), more than a third of the population was, so to speak, self-disenfranchised. In the last forty years, the highest level of participation was reached in 2004, for the presidential election pitting the incumbent George W. Bush against the challenger John Kerry. Despite the high stakes in these elections, only 56 percent of the voting-age population actually turned out and voted.

Apathy is not a recent phenomenon. Thomas Jefferson already bemoaned Americans' lack of political interest: "We in America do not have government by the majority. We have government by the majority who participate."[12] In *Democracy in America*, Tocqueville worried that in America, private interests and individualism often took precedence over participation in political affairs, a development that he feared would facilitate the development of despotism.[13] The quantitative methods of the twentieth century have made apathy tangible as a democratic plague, at least among old Western democracies.

The fact of apathy, however, need not support antidemocratic conclusions about the incompetence of an absentee public. For one thing, if apathy is a problem, it is not necessarily an unsolvable one. From the previously discussed rational choice perspective, apathy is in fact a "rational" behavior in large democracies, where the causal impact of any individual vote in a large election is essentially zero.[14] Decreasing the level of apathy might thus be a matter of changing incentives. Suitably modifying institutions so as to increase the incentives of citizens to vote and fostering more direct participation at different levels of the political process may help reverse the trend toward apathy. Ackerman and Fishkin (2004) have advocated financially remunerating people in order to entice them to participate in national "Deliberation Days." In other schemes, people might be rewarded for voting (if only symbolically, as is sometimes the case already, with a hot dog, a sticker, or a smile). In France, the French presidential candidate Ségolène Royal introduced proposals meant to include citizens at different levels of political decision making, with the goal of increasing citizens' sense of causal efficacy.[15]

---

[12] Although this quote is famous and seems fairly well established, I have yet to find the precise reference for it.

[13] Tocqueville thus wrote: "What is important to combat is therefore much less anarchy or despotism than the apathy that can create the one or the other almost indifferently." (Tocqueville 2002: 704)

[14] Referenda and initiatives are also rare, perhaps explaining citizens' feeling that their contribution does not really matter.

[15] She proposed the creation of citizens' juries selected by lot to assess the quality of conducted policies. She also constructed her political agenda and program based on suggestions made on her website, characteristically called a "participative forum."

The fact of apathy would not seem to challenge the idea of democratic competence, at least at the level of the group. Selective apathy at the level of the group may increase the group's capacities in decision making. People who choose to select out of the voting process may do so for good reasons. They might correctly identify that they do not know enough to vote. If voting is about making enlightened judgments, as opposed to merely expressing a preference or taking a guess, then it might be better for the least enlightened to refrain from expressing a poorly informed opinion. In the early 1950s, "elitist" theorists of democracy enunciated that position. In an article titled "In Defence of Apathy" (1954), H. W. Morris-Jones argued that apathy is a sign of political health and a potential guard against fanaticism.[16] In a similar vein, the comparative political sociologist Seymour Lipset condoned apathy because of what he deemed to be the more authoritarian inclinations of the uneducated and uninterested. To thwart totalitarianism, he thought, such characters were better left out of political decision making (Lipset 1960). The political philosopher Jason Brennan (2011) has more recently defended the comparable thesis that uninformed citizens are under a duty not to vote.

One may look at deliberation in the same light. Deliberation might be more likely to yield better results when the more informed speak up and the ignorant stay mum. Participation in either voting or deliberation would thus seem valuable from an epistemic perspective only to the extent that it contributes to raising the quality of the individual's contribution to the group decision.

While a defense of citizens' apathy on epistemic grounds may seem plausible, at least when applied to the more ignorant citizens, it assumes that the quality of the contribution of a given citizen to a deliberation or a vote can be predicted on the basis of his *ex ante* level of information. Further, this view also assumes that the mechanisms of deliberation and vote aggregation are unequipped to deal with the presence of wrong or ignorant input. As we will see later, there are several ways in which both assumptions can be questioned. In particular, we will see that while the correctness of an individual's contribution matters, the degree to which it introduces some amount of cognitive diversity in the collective vote or debate matters as well. Sometimes the benefits of the difference introduced by a mistaken or poorly informed view will outweigh the costs of the corresponding loss of accuracy. Thus voter apathy, even on the part of the ignorant and uninformed, can be a considerable evil for the epistemic

---

[16]Probably recalling the high levels of participation in Nazi Germany, he conjectured that "A State which has 'cured' apathy is likely to be a State in which too many people have fallen into the error of believing in the efficiency of political solutions for the problems of ordinary lives" (Morris-Jones 1954: 37).

quality of collective decisions. Apathy, including of the patently ignorant, is thus a double-edged sword that should not be celebrated without caution and should, in fact, often be deplored.

2.1.3 THIRD REPROACH: CITIZENS ARE IGNORANT

This is not to say that widespread citizens' ignorance is not a problem. On the contrary, civic illiteracy presents what is perhaps the more serious problem of all for democracy. This ignorance is now empirically well documented by six decades of public opinion research. To give just a few classic examples, half of the American public cannot name the two US senators from their own state, 20 percent of college graduates think Russia belongs to NATO, and a large majority of Americans do not know that Japan has a democratic system of government (Delli Carpini and Keeter 1996: 62). The citizens are so politically uninformed that their ignorance makes it difficult even to gather information about how little educated they are, as survey researchers fear asking too many factual questions might embarrass ignorant respondents and cause them to terminate the interview (Popkin 1994: 34).

More depressing still, the voters' lack of political knowledge has remained relatively constant across time. In the 1940s, just 36 percent of Americans could define the "welfare state." In 1952, just 27 percent of Americans were able to name two of the three branches of government, not including those who refused to respond to the question. In 1972, only 22 percent of Americans knew what Watergate was. More recent questions found that the majority of Americans are unable to locate Massachusetts on a map or to describe the New Deal; to the latter question, only 15 percent responded correctly. On this kind of question, contemporary Americans score only marginally better than Americans surveyed forty years ago. Many researchers conclude that most people know very little about politics, and that the statistical distribution behind that statement is fairly stable across time (e.g., Luskin 2002 and Converse 2000).[17]

These dispiriting results have been recently challenged by a literature on "the reasoning voter" (e.g., Page and Shapiro 1992; Popkin 1994; Lupia and McCubbins 1998). Part of the more general paradigm of "bounded rationality," which has brought to economics and political sciences the vocabulary of "heuristics" and "satisficing" as supplements and sometimes substitutes for that of "calculus" and "maximizing,"[18] this literature

---

[17] I have yet to find a reference to similar studies conducted in other democratic countries but there is no reason to think that the statement should not be true across cultures as well (even if absolute levels of knowledge may vary from one country to the next).

[18] Gerd Gigerenzer and others (2000, 2001) have developed a theory of rationality modifying or complicating (depending on the interpretations) the model of unconstrained rationality of economists. They follow in the wake of Herbert Simon's work (1957).

does not so much refute the fact that voters are little informed as it makes sense of this lack of political knowledge in terms of cognitive shortcuts and educated guesses. In this interpretation, voters are "cognitive misers," smartly and efficiently allocating their limited supply of brain cells and attention time so as to be able to vote appropriately with little knowledge. If they do not vote exactly how they would vote if they had gathered all the relevant information, they manage something close enough.

This defense of the reasoning voter makes convincing points and presents a remarkably intuitive idea of reason and rationality—at least, one more plausible than the perfect rationality postulated by rational choice theory. Yet it has had little influence on the perception that democracy is mostly about fairness and preferences, as opposed to competence and judgment. While Samuel Popkin urges that "we need a better theory of public knowledge about politics" (Popkin 1994: 36), it seems unlikely to happen until we take seriously the idea that democracy is not simply about preferences and desires but also about reason and judgment. In other words, a better theory of public knowledge about politics requires a move away from the politics of preferences toward the politics of judgment.

In a different vein, some people argue that the "public ignorance" argument invoked by theorists skeptical about democratic schemes premised on citizens' competence succeeds only in establishing that people are chronically misinformed, not that they are congenitally incompetent. This is, for example, Robert Talisse's reply to the objections raised against deliberative democracy voiced by Richard Posner (2002, 2003, 2004) and Ilya Somin (1998, 2004b). Posner and Somin argue that demonstrably high levels of public ignorance render deliberative democracy practically impossible, because citizens simply lack the cognitive abilities requisite for rational deliberation. According to Talisse (2004), however, the ignorance data invoked by these authors show that the public is ignorant in a way that does not necessarily defeat the requirements of deliberative democracy. Talisse thus draws a distinction between two concepts of ignorance—belief ignorance, or the fact of holding false beliefs for reasons external to the agent holding these beliefs (e.g., the existence of biased sources of information, or manipulation by political pundits), and agent ignorance, or the fact of holding wrong beliefs as a result of a cognitive failure on the part of the individual herself (mistaken inferences and/or carelessness) (pp. 455–56). In the first case, the person is more accurately called "misinformed." Only in the second case does the person deserve to be called ignorant and, indeed, "in cases where the cognitive failure is particularly egregious, . . . incompetent" (p. 458). According to Talisse, the literature appealed to by Posner, Somin, and others to diagnose general civic incompetence in the citizenry only demonstrates a high degree of belief ignorance but does not succeed in demonstrating agent ignorance. This is lucky, as Talisse concludes, since if agent ignorance was established for a fact, it would prove "devastating to every conception of democracy, not just the deliberative one" (p. 461). As things stand, however, Talisse argues, advocates of deliberative democracy can always appeal to the necessity of repairing the civic institutions responsible for a high degree of citizens' ignorance, without having to grant citizen incompetence (p. 459).

Empirical evidence, however, can be interpreted as supporting the view that people not only are ignorant but cannot think coherently either.[19] Philip Converse's influential study "The Nature of Belief Systems in Mass Publics" (1964) thus determined that Americans both hold incoherent political views and do not think coherently about political issues.[20] In particular, most of Converse's subjects seemed to harbor contradictory and constantly changing opinions about most policies. For instance, 42 percent of Converse's respondents evaluated policies strictly by how much they thought that certain groups would benefit from their implementation. Such respondents often supported incongruous sets of policies (such as tax cuts and increased social spending simultaneously) if both were perceived as beneficial to some group (p. 216). Meanwhile, another 24 percent were placed in the "nature of the times category," in which "parties or candidates were praised or blamed primarily because of their temporal association in the past with broad societal states of war or peace, prosperity or depression" (p. 217). Unable to discern the impact of specific policies, this group merely assumed that the party or politicians in power are responsible for whatever good or bad befell the American political landscape. Finally, in 22 percent of respondents, Converse was unable to discern any "issue content" whatsoever.

All in all, the level of ignorance and incoherence that has been observed in citizens challenges the notion of democratic competence. If democracy is, indeed, the rule of such apathetic, ignorant, and incoherent citizens, how credible as a regime can democracy really be? From a purely prudential or instrumental point of view, why would we want to resort to democratic procedures in the first place?

2.1.4 FOURTH REPROACH: DEMOCRATIC DECISIONS ARE IMPOSSIBLE AND MEANINGLESS

If we now turn away from the individual to the group, critics of democratic decision making turn out to be just as harsh. William Riker's

---

[19]That is, in Talisse's terminology, they are not just belief ignorant, but agent ignorant as well.

[20]Note that this method of study does not just test how much subjects know but also how much they think about political issues in order to develop reasonable opinions.

Liberalism against Populism (1982) has convinced generations of political scientists that democratic decision making in itself is impossible, arbitrary, and meaningless. Riker develops two main arguments against what he calls the "populist" interpretation of democracy. The first relies on Arrow's (1953) Impossibility Theorem to conclude that there is no such thing as a "general will." According to Arrow's theorem, no voting system based on ranked preferences can meet a certain set of reasonable criteria when there are three or more options to choose from. These criteria are unrestricted domain, nonimposition, nondictatorship, monotonicity, and independence of irrelevant alternatives. Arrow's theorem is itself a generalization of the Condorcet paradox which refers to the possibility of cycling in elections.

The other argument proposed by Riker is that once we have chosen a social decision mechanism, the results may not represent the true wishes of the population, since people may vote strategically.[21] To make that claim, Riker relies on a lesser-known theorem, the Gibbard-Satterthwaite theorem (1972), which shows that any voting method that is completely strategy-proof must be either dictatorial or nondeterministic. One constraint that a social choice function must satisfy in the framework of the Gibbard-Satterthwaite theorem is the same "unlimited domain" of preferences that is also present in the framework of Arrow's theorem.

The normative claims about democracy that Riker derives from the two theorems have been challenged both theoretically and empirically. Empirically, it can be argued that problems of cycling, agenda control, strategic voting, and dimensional manipulation are not sufficiently harmful, frequent, or irremediable to be of normative concern (e.g., Mackie 2003).[22] Theoretically, all the assumptions that the theorems depend on can or have been criticized, perhaps most especially the assumption of "unlimited" or "unrestricted" domain from which preferences are supposed to be taken.[23] It has been argued that no matter how troubling

---

[21] Strategic voting (also called tactical or sophisticated voting) occurs when a voter supports a candidate other than his or her sincere preference in order to prevent an undesirable outcome.

[22] Mackie dismisses as erroneous the historical examples adduced by Riker to illustrate the problems of cycling and instability (in particular Riker's argument that the US Civil War was caused by arbitrary dimensional manipulation).

[23] Contrary to a common misinterpretation, this assumption does not mean that the set of feasible alternatives must include any possible desire or preference of any citizen, so that it would be enough to solve Arrow's theorem to ban some undesirable preferences (say, for racist policies). It simply means that once a feasible set of alternatives is specified then any pairwise ordering (essentially what a preference is in this context) of that feasible set must be admissible. So racist policies could be excluded from the feasible set; and yet, so long as we have a set of at least three remaining feasible options where any pairwise ordering is admissible, Arrow's theorem would apply.

instability theorems may be, they do not justify Riker's pessimist conclusion. These theorems simply "demonstrate the importance of gaining a fuller understanding of the likely performance of democratic institutions" (Coleman and Ferejohn 1986: 25).

I will consider in greater detail in chapter 7 attempts made by critics to take the bite out of social choice theory results. The important point is that in spite of these various refutations, the idea remains profoundly influential in democratic theory that there is no such thing as a politically coherent democratic decision. Riker and his followers have entrenched the idea that democracy is inefficient and meaningless and that, in effect, "the only voting method that isn't flawed is a dictatorship" (according to a common and slightly tendentious restatement of Arrow's theorem).

The so-called positive side of democratic theory, which describes democratic incompetence at both the individual and collective level, fosters rather antidemocratic conclusions. If we now turn to the more recent normative approaches, we will see that the content of these theories is rarely premised on the idea that the people are competent to rule or that they find ways to account for democracy's performance in ways that give all the credit to the more aristocratic elements of representative democracy. Elitist theories of democracy, in particular, are explicitly designed to circumvent the problem of regular citizens' competence. As to participatory and deliberative theorists of democracy, they mostly steer clear of competence issues and outcome-oriented considerations, focusing instead on the intrinsic value of democratic procedures. The only exception is a tiny but growing group of "epistemic democrats."

2.2 Normative Approaches

2.2.1 ELITIST THEORIES OF DEMOCRACY

On the more normative side of democratic theory, one finds first an ambiguously democratic stance amongst Schumpeterian or so-called elitist theorists of democracy (of which William Riker, mentioned above, is a representative). Elitist theorists do not deny that politics requires judgment and competence to make intelligent decisions. They simply do not believe that this competence is to be found in the people. A good illustration of this lack of trust in the people's political capacities is Schumpeter's disparaging remarks about democratic citizens, whether as individuals or as a group. Regarding individuals, Schumpeter argues that the typical citizen "drops down to a lower level of mental performance as soon as he enters the political field. He argues and analyzes in a way which he would readily recognize as infantile within the sphere of his real interests. He becomes a primitive again" (Schumpeter [1942] 1975: 262). Regarding large groups of citizens, Schumpeter argues, following Pareto and Gustave Le Bon, that they are characterized by "a reduced sense of responsibility, a lower level of energy of thought and greater sensitiveness to non-logical influence" (p. 257).

Schumpeter does not simply criticize the multitude in the sense of a physical agglomeration of many people. He also criticizes "newspaper readers, radio audiences, members of a party," who, in his view, "even if not physically gathered together are terribly easy to work up into a psychological crowd and into a state of frenzy in which attempt at rational argument only spurs the animal spirits" (p. 257). While Schumpeter believes that newspaper readers, radio audiences, and members of a party are terribly easy to manipulate, he also does not trust those citizens who vote without reading the newspapers, listening to the radio, or caring about party affiliation. Schumpeter is not Rousseau. Schumpeter simply believes that the rational thinking-through of politics should be left to elites. For him, elites consist of those individuals capable of resisting the laws of group psychology, or forming a sufficiently small group as to make those laws inapplicable.[24]

Elitist conceptions of democracy circumvent by definitional fiat the problem of the people's competence. Citizens are turned from political agents into political consumers (a transformation completed by the publication of Anthony Downs's "economic theory of democracy" in 1957). The irony, Finley remarks, is that while the twentieth century undeniably brought much disillusionment about the wisdom of majorities, especially on the Continent, elitist theories of democracy became most influential in the very countries—Great Britain and the United States—where democracy had been most empirically successful.

Elitist theories thus justify democracy *in spite of* the alleged incompetence of the average voter. An elite of professional politicians is required to take the reins of the state, and the people's role is limited to assessing the quality of the politicians' performance by reelecting them or kicking them out of office at the end of their mandates. The capacity of the people to validate retrospectively or punish their leader's decisions forms a post hoc accountability mechanism giving leaders an incentive to act in the collective interest. The elitist reading of politics is clearly cognitive/epistemic: rulers are assumed to *know* what the common good is, even though they may lose the incentive to pursue it, which is why a minimal democratic control through periodic elections is required.

As already said, in elitist theories of democracy, the interpretation of voters' apathy is not necessarily negative. For Schumpeter, citizens should have as little to do with political decisions as possible, their role ideally

---

[24]Notice the similarity with Tocqueville on that point. Contrasting the advantages of the rule of an elite few with those of a democracy and a kingship, Tocqueville remarks that "an aristocratic body is too numerous to be captured, too small in number to yield readily to the intoxication of unreflective passions. An aristocratic body is a firm and enlightened man who does not die" (Tocqueville 2002: 220).

being limited to selecting competing leaders. Schumpeter even opposed the idea of citizens' contacting their representatives, writing letters to their senators, and engaging in political activities beyond the act of voting. Schumpeter describes a model in which widespread participation is unessential—one in which apathy can exist, yet democracy can survive. To some degree, apathy on a large scale is even a good thing for Schumpeter, provided it is not so overwhelming as to render elections superfluous and make the competitive selection of leaders impossible.

2.2.2 PARTICIPATORY THEORIES OF DEMOCRACY

Unlike elitist democrats, participatory democrats are committed to the notion that common people should have a say in political matters. They also believe in the educative effects of participation on the average citizen. The more one takes part in debates about means and ends at the local or the national level, in the political and economic sphere, or even the private sphere of the family, the theory runs, the more likely one is to develop the qualities of an enlightened and competent citizen and also the confidence that one is such an enlightened and competent citizen (e.g., Pateman 1970; Fung and Wright 2003). Participatory democrats, however, generally do not justify democracy only in the name of this educative or empowering property of participation. Instead, participatory democrats more often argue, democracy is good insofar as it expresses the citizens' "autonomy" or freedom to choose (e.g., Barber 1985) or principles of equity and efficiency (e.g., Fung and Wright 2003). But even if and when they try to rely on the educative properties of citizens' participation to justify democracy as valuable, one can argue that to the extent that they consider enlightenment of the masses a mere byproduct of participation, as opposed to its direct consequence or the goal of democratic decision making, such a justification remains ambiguous and weak (for the limits of a justification by the by-product of a practice, see Elster 1989a). This ambiguity arguably points to a level of uneasiness that participatory democrats have with the idea that the people could ever be wise, or at least with the principle of resting the case for democracy on citizens' wisdom.

2.2.3 DELIBERATIVE THEORIES OF DEMOCRACY

Most surprisingly, perhaps, many theorists of deliberative democracy have been for a long time, and to a large extent still are, reluctant to phrase their justification of democracy and, specifically, democratic deliberation among regular citizens in terms of expected epistemic benefits. This reluctance is probably due to the fact that deliberative democracy is a family of views concerned with providing a theory of the legitimacy of democratic decision making based on the centrality of "public deliberation of free and equal citizens" (Bohman 1998: 401), but one that is originally noninstrumental—that is, unconcerned with the quality of outcomes that can be expected from democratic procedures—and certainly not consequentialist. In other words, deliberative democrats see democratic legitimacy "in terms of the ability or opportunity to participate in effective deliberation on the part of those subjects to collective decisions" (Dryzek 2000: 1), not in terms of the benefits to be expected of democratic deliberation for these subjects. Deliberative democrats' definition of democratic legitimacy thus implies that the legitimacy of political decision making and self-government derives from deliberation as a procedure per se, and the values it embodies, not from the expected results of deliberation. There seems to be something intrinsically good about involving the people in direct or indirect deliberation, regardless of the epistemic properties that can be expected of the latter. The idea is that a deliberative procedure that allows everyone a say in the final decision is good because including everyone expresses values such as autonomy, equal consideration and respect, or political equality. As David Estlund (2008) observes, this position, an example of pure (or fair) proceduralism, can be attributed to a number of contemporary democratic theorists (e.g., Manin 1987: 352–59; Sunstein 1988; Cohen 1989; Elster 1995; Gutmann and Thompson 1996, 2004; Waldron 2001; Pettit 2003).

In many deliberative democrats' view, the people have a right to participate in decision-producing deliberation because deliberation is a fair decision procedure, not particularly because they are competent decision makers or because deliberation tends to yield good results. In other words, deliberative democrats tend to dissociate entirely the question of people's right to self-rule from the question of their competence for it. Deliberative democrats may care about the quality of deliberative outcomes, but it does not seem to enter their conception of democratically legitimate deliberations or deliberatively produced decisions.

Most deliberative democrats fall into that proceduralist category. Those who do build in substantive requirements, such as Joshua Cohen (1996) or Amy Gutmann and Dennis Thompson (2002, 2004: especially chap. 3), phrase the substantive constraints in terms of "rights." The respect for rights may count as a substantive standard by which to assess the intelligence of democratic outcomes, but it is not one that need be part of a conception of democracy per se. It could simply be part of a liberal framework grafted on top of a theory of deliberative democracy and constraining from without. Furthermore, neither Cohen—though one of the first to theorize the notion of epistemic democracy (1986)—nor Guttmann and Thompson, nor some other prominent deliberative democrats ever fully and explicitly commit to an epistemic approach to democracy (see in particular Cooke 2000 for an analysis of the ambiguities in the work of Joshua Cohen and Seyla Benhabib).

2.2.4 EPISTEMIC THEORIES

In recent years, however, a few democratic theorists have started to take an interest in the epistemic value of the decisions made following a democratic procedure. I will only mention here those authors who identify themselves as epistemic democrats, although more could potentially be characterized as such: David Estlund (1993a and b, 1994, 1997, 2002, 2007), Carlos Niño (1996), Thomas Christiano (1996, 1997), Gerald Gaus (1996, 1997a and b), Robert Goodin (2003 and 2008), Josiah Ober (2010, 2012), Robert Talisse (2009), and Fabienne Peter (2009). Habermas may also be invoked as, in a recent article (2006), he seems to have no qualms being categorized as an "epistemic democrat," arguing forcefully in particular for the empirical plausibility of the epistemic properties of deliberation.[25] There is thus a growing number of democratic theorists, and specifically among them, deliberative democrats, who are arguing for democracy on the basis of the epistemic properties of some of its procedures or institutions.

Put generally, in the epistemic interpretation, democratic decision-making processes are valued at least in part for their knowledge-producing potential and defended in relation to this (Peter 2009: 110). Epistemic democracy further combines deliberative and aggregative approaches to democracy but shifts their focus toward an outcome-oriented consideration for how well democratic procedures such as deliberation and voting help democratic decisions approximate a procedure-independent standard of correctness. The assumption of a "procedure-independent standard of correctness" is, indeed, one of the key characteristics of an epistemic approach to democracy, whether the democratic procedure referred to is voting (see in particular Cohen 1986 for the clearest conceptualization of what an epistemic conception of voting means) or deliberation. The procedure-independent standard is often referred to as the

---

[25] This epistemic stance is somewhat surprising since Habermas articulated in earlier writings a distinction between the moral and political-legal sphere and asserted being a cognitivist only in the moral sphere, while remaining a proceduralist in the political-legal sphere. To the extent that Habermas defends an ideal deliberative procedure as forming the independent standard of assessment of actual political decision procedures and their outcomes, however, he does qualify as an epistemic democrat. On that reading, the political-legal procedures are themselves dependent on the truth value of some moral claims. If Habermas qualifies as an epistemic democrat, one might argue that so would Rawls (who also proposes the ideal procedure of an original position as the benchmark of constitutional political decisions). However, since Rawls claimed "epistemic" abstinence, it seems difficult to pigeonhole him forcibly in a category he explicitly rejects. I say more on the compatibility of political cognitivism and epistemic abstinence at the end of chapter 8.

general will, the general interest, the common good, truth, the right, the good, or any other term indicating that we are after a certain type of ideal outcome that reflects the best interest of the people and is independent from their actual, possibly mistaken preferences, beliefs, or judgments.[26] Specifically, for epistemic democrats, the outcome of a vote or a deliberation is not, as for a pure proceduralist, defining the common good but, at best, fallible evidence for it.

While epistemic democrats remain a minority, it has been argued (by epistemic democrats, essentially) that deliberative democrats subscribe, knowingly or not, to the assumption of an independent standard of correctness, at least on some issues, and, therefore, an at least partially epistemic conception of democracy. This is the case made, for example, by both José Luis Martí (2006) and David Estlund (2008, specifically chap. 5). As Martí puts it, a commitment to the assumption of a procedure-independent standard of correctness for deliberative outcomes is largely unavoidable, since

> as a discursive process based on reason, deliberation assumes . . . both *the existence of rightness* (or impartiality, or some other equivalent) in political decisions, and *the possibility of knowing which is the right* (or impartial) decision. . . . To argue in favour of decision A means, briefly, to show that decision A is *the right* decision, or at least, that A is *better in terms of rightness* than other decisions being compared. (Martí 2006: 29; emphasis in original)

According to Martí, participants in deliberation "assume"—that is, underwrite a commitment to—both the existence of "some intersubjective criterion of validity for their claims" and the possibility of knowing it (at least to some degree). This criterion needs to be "*at least partly independent* from the participants' preferences, desires, or beliefs, and from the process itself" (Martí 2006: 30, my emphasis).[27]

David Estlund (2008) similarly establishes how "the flight from substance" of what he calls "deep deliberative democracy" ultimately fails (chap. 5). Willingly or not, deliberative democrats must posit something outside of the actual deliberative procedure by which they assess its outcome, whether it is an ideal outcome or range of outcomes or an ideal procedure. Estlund further shows how so-called fair proceduralists, such as Jeremy Waldron, posit an independent standard of correctness (an ideal fair procedure). Even social choice theorists, who deny that there is a common good but subscribe to an ideal procedure of aggregation,

---

[26] See chapter 8 for a deeper analysis of this notion of procedure-independent standard of correctness.

[27] See also Cohen 1986: 34ff.; Estlund 1993a: 74, 79–81; and Estlund 1993b: 1448ff..

are closer than they think to endorsing an epistemic approach to democracy. The assumption of an independent standard of correctness of some kind, and the possibility of knowing it, thus undergirds most theories of democracy, including those that deny it. Even though it has yet to be recognized as a mainstream paradigm, epistemic democracy thus arguably marks a turning point in democratic theory, away from pure proceduralism and toward a greater consideration for democratic outcomes and citizens' competence.

The questions that epistemic democrats address are of two kinds. Some have to do with the normative authority of democracy and whether or not this normative authority, defined as a right to rule (possibly correlated with a duty to obey), includes an epistemic dimension in addition to the classically included dimensions of fairness, consent, and procedural justice. This is the kind of question Estlund, most prominently, addresses in his defense of a new "philosophical framework" for the normative concept of democratic authority, one that takes seriously the question of democratic epistemic competence and makes the question of a right to self-rule partially conditional on it (Estlund 1997, 2008). Another set of questions considers whether democracy has, indeed, any epistemic properties and, if so, in what amount and through which mechanisms. Let me say a few words about what has been accomplished so far with respect to each type of question.

2.2.4.1 Establishing the Epistemic Dimension of Democratic Authority

In terms of supporting the epistemic dimension of democratic authority, the milestone is David Estlund's book *Democratic Authority: A Philosophical Framework* (2008). Besides being the most important epistemic account of democratic authority to date, it articulates different parameters of political legitimacy in a novel and interesting way. One of the main arguments of the book is that purely proceduralist (or, as he calls them, deeply proceduralist) justifications for democracy fail. If the fairness of the procedure was all that mattered in deciding between two political proposals, then flipping a coin would be just as good as counting votes, since in both cases all citizens have an equal (fair) chance of influencing the outcome.[28] After thus establishing the necessity of an epistemic account of the normative authority of democratic decision procedures, Estlund then proceeds to propose his own theoretical account of democratic authority, which he calls "epistemic proceduralism." According to epistemic proceduralism, democratic decisions are legitimate insofar as

---

[28] Close to zero in the case of voting, or exactly zero in the case where we flip a coin.

the democratic procedures that yielded those democratic decisions have
a better-than-random tendency to produce right or correct decisions.
This approach is basically a reintroduction of a form of instrumentalism
into normative justifications of democracy (which is not to say a form
of consequentialism, as Estlund explicitly rejects that position (2008:
164–65)). Democracy can only be a legitimate and authoritative decision
procedure if it can be expected to meet a minimal threshold of epistemic
competence, and not just because of its procedural fairness or the values
it expresses. Another way to say this is that Estlund reconnects the question of the justification of democracy and that of its legitimacy.[29]

This is not to say that Estlund's approach is purely instrumental or
makes no room for the principle of consent. What makes the general theory of democratic authority proposed by Estlund fall out of the category
of purely instrumental justifications of political authority is the fact that
epistemic proceduralism is just one part of the theory. Two other parts of
the theory are the general acceptability requirement and a theory of "normative consent," both of which make room for noninstrumental reasons.
The "proceduralism" in "epistemic proceduralism" refers to these other
parts of the theory.

I will say a bit more about the problems raised by this partly procedural, partly instrumental approach to democratic authority at the end
of this chapter, but for now let me remark that while I admire Estlund's
attempt to demonstrate the epistemic dimension of democratic authority and find myself largely persuaded by it, in this book I wish to avoid
entirely the discussion of the nature of normative political authority and
the question of whether epistemic reliability is a part of it or not. The
question of political authority is a maze that I myself do not wish, or
need, to enter here.

What interests me is simply to show that whatever intrinsic fairness
democratic procedures may have, and whatever the sources of democratic
authority may be, democratic procedures can be shown to have epistemic
qualities and that those provide powerful reasons to value democratic
procedures and embrace the outcomes they produce. Whether epistemic
properties add to the legitimacy of democratic decisions in general or
simply provide prudential reasons to abide by them is a question I will
thus leave unaddressed.

---

[29] A connection severed—or rather, entirely denied—by consent theorists who locate
legitimacy merely in the actual, tacit or explicit, consent of the governed (e.g., Simmons
2001) or pure proceduralists, including many deliberative democrats and "reasonable consensus" theorists like Rawls, for whom the legitimacy of the decision is simply determined
by the fairness or some other intrinsic property of the decision procedure.

2.2.4.2 Demonstrating that Democracy Has Epistemic Properties

The only questions of relevance here will thus be the second type of question addressed by epistemic democrats: Does democracy meet a minimal threshold of competence? Can it in fact be expected to do better than this? How does it fare, in particular, compared to allegedly more promising alternatives like an oligarchy of the wisest?

On these specific questions, Estlund's work does not provide a compelling answer. While his book may succeed in its normative project—namely establishing that epistemic reliability is a necessary component of democratic authority—it is less compelling at fulfilling the corresponding positive task: that of actually giving us reasons to believe that democracy is epistemically competent. What is arguably lacking is a clear analysis of the mechanisms accounting for the presumed epistemic reliability of democratic procedures.[30] Estlund does briefly consider two possible "epistemic engines"—the Condorcet Jury Theorem and what he calls the "democracy/contractualism analogy"—but ends up distancing himself from both. In the end, the credit is simply attributed to deliberation in virtue of its alleged information-pooling properties.

While Estlund may ultimately be right in his suggestion that democracy's epistemic competence, if it has any, owes more to its deliberative practices than to its aggregative practices, one may object to the rejection of the aggregative account altogether. There are, after all, other explanations of the epistemic properties of vote aggregation beyond the Condorcet Jury Theorem, the most important of which will be covered in the central chapters of this book. Furthermore, while Estlund's ultimate reliance on deliberation as the main epistemic engine of democracy is intuitive and convincing, it is also so broadly tailored as to serve not just democracy but any collective decision procedure that is minimally deliberative, say, a deliberative oligarchy of the smartest people in the group. What is required, in order to support the case for democracy as opposed to any other collective deliberative rule, is an argument showing why deliberation needs to be democratic and specifically inclusive of all members of the group in order to have epistemic properties superior to those of a random decision procedure. In that respect, Estlund's book is an invitation to pick up where he left off: at the threshold of the idea of collective intelligence applied to the justification of democracy.

A few other epistemic democrats besides Estlund have addressed the question of the epistemic properties of democracy. Elizabeth Anderson (2006) has thus put forward a deliberative, essentially Deweyan argument in favor of the epistemic properties of democracy. On the more aggregative

---

[30] This is particularly clear in Estlund's chapter 9—"How Would Democracy Know?"—which never fully answers the question it raises.

side, Robert Goodin has proposed an interpretation of democracy as a "Condorcet truth tracker" and a "Bayesian persuader" (2003: chaps. 5 and 6) and addressed from a partially empirical perspective the issue of the epistemic properties of deliberation (2008). Josiah Ober (2010), for his part, has provided the most comprehensive and innovative epistemic analysis of direct democracy as exemplified in ancient Athens. He argues that the superiority of the democratic city-state over its oligarchic rivals can be credited to the ability of Athenian institutions to pool information and knowledge most efficiently. Robert Talisse (2009), finally, from a more philosophical point of view, has argued for democracy as the only way to solve moral conflicts in epistemically satisfying ways.

As of now, however, existing epistemic arguments for democracy are piecemeal and do not, in particular, amount to a systematic case for democracy's superiority over alternative regimes. This perhaps explains why it is that "while supporters of epistemic democracy have explicitly defended this conception, other deliberative democrats have simply not mentioned the epistemic case for democracy" (Martí 2006: 28). This is not a failure of the approach per se but goes to show that epistemic democracy and its advocates still need to gain in credibility. More work is needed on the specific epistemic mechanisms of democracy and the way they supplement each other, opening many avenues for new research in democratic theory.

Such work is important because as long as democracy is interpreted as the rule of the dumb many, it seems hardly justifiable (if not legitimate) as a collective decision rule. Regardless of the empirical success of actual democracies—which, for all we know, might well have succeeded only to the extent that they were not in fact full democracies, or just due to sheer luck, given a serendipitous historical and geopolitical environment—the question of citizens' competence and the quality of democratic outcomes poses a challenge to the value of democracy.

There are essentially two democratic ways to answer this question. One consists in denying the existence of a tension between the belief in democracy's value and the belief in citizens' incompetence by rejecting the premise that the people are as uninformed, unengaged, and irrational as the literature on public opinion makes them out to be. This way is encumbered by an accumulation of empirical evidence difficult to challenge. Following this strategy implies reinterpreting a lot of empirical evidence as misguided measurement of "political intelligence" and offering an alternative account of rationality and what is sufficient for voters to make competent decisions. The literature on the rational voter, for example, goes a long way in this direction, reinterpreting voters' lack of information and seeming apathy as rational cognitive parsimony offset by the resort to informational shortcuts and other "heuristics" (e.g., Popkin 1994).

Another alternative consists in pursuing another argumentative line: the idea that democracy is actually the better decision-making procedure overall, thanks to the phenomenon of collective intelligence. In that view, a comparison between rule of one and rule of the many would actually yield a democratic conclusion and provide an independent epistemic reason to favor democracy over oligarchy. Taking the idea of collective intelligence seriously, I conjecture that in politics there might not be a better "knower" than the group itself, which under the right conditions can outsmart anyone within the group, including the smartest members.

In other words, on the view defended here, democracy is a political system capable of turning the input of not-so-smart citizens into relatively intelligent decisions. My position thus consists (1) in admitting that the question of the competence of the people and of the quality of democratic outcomes is relevant to the question of democratic justification and (2) in arguing that this is not, however, a problem, since the people are actually competent, if not as individuals, then at least as a group, and if not on every issue, then at least overall.

We will soon see that such trust in the people and, more generally, democratic rule has an interesting genealogy. Before moving on to the next chapter, let me address a final worry that may linger on the reader's mind. Isn't there a risk, in reintroducing consideration for competence and outcomes into our justification of democracy, of justifying an oligarchy of the wise in the end?

2.2.4.3 The Risk of Epistocracy

How far can we go in taking an instrumental perspective on the justification of a decision rule? What weight, in particular, should sensitivity to outcomes have in comparison with other reasons we have to value democracy? If we put more weight on epistemic considerations beyond the minimal threshold of "better than random," do we not risk justifying not just democracy but any alternative that does better on epistemic grounds? In the same way that opening up the possibility of discussing the epistemic competence of the voter may invite restrictions on the franchise, assuming that a certain amount of epistemic success is necessary for a political regime to have any value at all may suggest delegation of political choices to a caste of "knowers" or, as Estlund calls them, "epistocrats" (Estlund 1997:183).

This objection can be answered in several ways. First, there might be reasons other than epistemic ones to involve the people in the decision process—for example, considerations of fairness and equality, or considerations for the way in which the procedure increases citizens' knowledge or improves their sense of belonging to the community. Charles Beitz makes this argument when he criticizes best-outcomes accounts of political authority for failing to take into account the side benefits of certain procedures (Beitz 1990: chap. 2). Bringing up competing reasons to value democracy is satisfying, however, only to the extent that we assume that democracy at least does better than random. When that threshold is met, but only then, considerations other than the epistemic can enter in the picture and take precedence. On some specific questions, though, or in some contexts, we might value the quality of results over the way they are achieved. Where experts prove smarter than the people, a consideration for epistemic competence will support antidemocratic conclusions for these questions.

One theoretical move blocking this conclusion would consist in denying that the identification of knowers is possible in politics. If it is impossible to identify universal political knowers of the kind Plato envisioned—people who are systematically better political decision makers than others—then democracy, a decision rule that involves everyone with an equal claim to knowledge (or lack thereof), becomes the default option. The problem is that there exists de facto consensus on political experts: pundits, academics, professional politicians, intellectuals courted by TVs and newspapers. Obviously, it is not hard to identify people who know more than others on political topics.

David Estlund's solution consists in denying that claims to rule of even acknowledged experts could be beyond the reasonable objections of individual citizens. In Estlund's theory, epistemic reliability simply marks a necessary threshold that a legitimate regime must meet. Beyond that threshold, however, other factors matter, which a priori rule out dictatorship or rule of the few or any other form of epistocracy, no matter how much more epistemically competent these alternatives can be shown or assumed to be. Estlund's blocking move, however, is premised on a liberal conception of a "qualified acceptability requirement" which is not without difficulty. Estlund (1997: 183) considers that because citizens should refuse to surrender their moral judgment on important matters to anyone, even if there might be a standard of correctness and even knowers of various degrees, there will never be any moral basis for epistocracy.

The problem is that unless one accepts Estlund's standard of a "qualified acceptability requirement," his theory remains vulnerable to the challenge of epistocracy. Estlund asserts that his theory of an acceptability requirement is "the only way to answer" the challenge of epistocracy, "since it is certain that there are subsets of citizens that are wiser than the group as a whole" (2008: 40). This commitment to the view that democracy is bound to be less epistemically reliable than epistocracy characterizes Estlund's self-proclaimed "epistemic modesty." This epistemic modesty, in my view, concedes too much and is, furthermore, unwarranted.

One could argue that a better answer to the challenge of epistocracy consists in arguing that it is not at all certain that there are subsets of citizens that are wiser than the group as a whole, at least if that statement is meant to be true as a general rule, rather than to describe occasional and topical exceptions. As the rest of this book will show, for most political questions and all things being equal otherwise, we have good reasons to believe that the group is likely to be on average at least as smart as any subset of its members. My solution thus consists in arguing that epistocracy is not a tempting option, not because we can never agree on who the knowers are, as sometimes we do, or because there could and should never be a consensus among qualified or reasonable points of view on who they are, as this claim relies on debatable normative assumptions, but because there are reasons to think that the more reliable knower is actually the group as a whole, as opposed to any particular individual or group of individuals within it.

CHAPTER THREE

A Selective Genealogy of the Epistemic Argument for Democracy

WHILE THE SALIENCE OF THE IDEA of collective intelligence as an argument for democracy is recent, it has an old and prestigious pedigree, which can be traced all the way back to the Sophists. This chapter aims to provide the historical background of the epistemic case for democracy by identifying its origins in the arguments of a few prominent thinkers in the Western tradition. As a necessarily cursory survey of proto-epistemic arguments for democracy, its goals are threefold. First, this chapter aims to bolster the intuitive appeal of the idea of collective intelligence applied to democracy by showing that this idea has a long history and quite a few ancestors. Second, this selective genealogy will allow me to identify the main mechanisms presented by the tradition as key to the phenomenon of collective intelligence. I say more on this below. The third goal of this chapter is to show—through a chronological approach—the evolution and limits of preexisting accounts of the phenomenon of collective intelligence in politics and pave the way for the more fine-grained and in-depth approach offered in the following chapters, on the basis of recent advances in the social sciences.

To anticipate some of the conclusions, we will see that the forerunners of the epistemic argument for democracy fall roughly into two camps: the "talkers" and the "counters."[1] Both arguably find their origins in the myth reported by the Sophist Protagoras in Plato's eponymous dialogue. According to the myth of Protagoras, every human being is endowed with a parcel of political knowledge. A first tradition—running from Aristotle to Dewey—develops that idea in a deliberative direction. The contemporary heirs of such a tradition are the deliberative epistemic democrats (e.g., Cohen, Estlund, Habermas). This deliberative tradition suggests that if the many can be smarter than, or at least as smart as, the few, it is because democratic discussion mirrors at the level of the group the individual element of reason present in each citizen. In that view, democratic reason is a function of individual reason; democratic reason is, indeed, individual reason writ large.

---

[1] My thanks to Jacob Levy for suggesting this dichotomy.

The other tradition, unfolding much later, takes up the aggregative idea also present in the myth of Protagoras. Spinoza is possibly the first to refer to the advantage conferred by the sheer number of people involved in the making of a decision. This tradition focuses on the epistemic properties of judgment aggregation when large numbers of people are involved. This intuitive idea takes proper analytical shape in the Jury Theorem of the philosopher and mathematician Condorcet, which puts the law of large numbers in the service of an epistemic argument for majority rule. Unlike the deliberative tradition, the aggregative tradition is heavily dependent on the progress of the mathematical science and, in particular, probability theory.

Rather than group authors according to the mechanism that is most central to their account, I follow in this chapter a chronological order that will take us from the Greeks to the first half of the twentieth century. This chronological order seems the most appropriate since some authors, particularly later ones like Mill or Hayek, do not even fall neatly in the categories of either "counters" or "talkers"—in Mill's case because he believes in both deliberation and the epistemic properties of an unruly "market of ideas," and in Hayek's case because his defense of the epistemic properties of markets turns out to be largely orthogonal to the question of the best government.

Before proceeding, let me add a caveat as to the nature of the history of ideas attempted here. This chapter does not offer the kind of in-depth and exhaustive textual exegesis that usually supports claims about what authors from the past really meant. I use selected passages, chosen on the basis of their greater relevance or representativeness, and extract from them what I take to be plausible in order to suggest that they might be construed as proto-epistemic arguments. In that sense, I am more interested in identifying in these texts the possible origin of contemporary arguments than in what their authors may or may not have literally meant to say (although I try as much as possible to respect the original intent). A careful history of the epistemic argument for democracy would probably demand, and be worth, a book-length study. The following survey merely attempts to pick up on and trace out key threads that are suggestive of the historical and philosophical origins of the epistemic case for democracy; I do not pretend to weave the different threads into a unified pattern.

Another limit of this survey, finally, is that it only discusses Western political thought. As a nod to other traditions, I would like to mention a few remarks by one contemporary Muslim political thinker, Muhammad Asad, who finds something very much like the idea of "collective intelligence" in the Qur'an itself. Reflecting on the rationale for majority rule, which he contrasts with the rule of the amir, Asad observes:

[t]he best we can hope for is that when an assembly composed of reasonable persons discusses a problem, the majority of them will finally agree upon a decision which in all probability will be right. It is for this reason that the Prophet strongly and on many occasions admonished the Muslims: "Follow the largest group" and "It is your duty to stand by the united community and the majority." (Asad 1980: 49–50)

According to Asad, for the Muslim tradition, the mechanisms through which the group is more likely to find the right answer to a given collective problem seem to be, as in the Western tradition, deliberation and majority rule. Asad further remarks:

In fact, human ingenuity has not evolved a better method for corporate decision than the majority principle. No doubt, a majority can err; but so can a minority. From whatever angle we view the matter, the fallibility of the human mind makes the committing of errors an inescapable factor of human life; and so we have no choice but to learn through trial and error and subsequent correction. (Ibid.)

Asad here embraces deliberation followed by majority rule as a fallible but overall reliable way to make collective decisions. I take this conclusion to be a striking indication that the proto-epistemic arguments in favor of deliberation and majority rule surveyed in the following have meaning beyond Western culture and are perhaps, indeed, of universal value. As to Asad's mention of the possibility of "subsequent correction," I take it to refer to the dynamic process of learning over time that is a part of democratic reason, which this book will discuss only briefly by way of a conclusion.

1. The Myth of Protagoras: Universal Political Wisdom

The first proto-epistemic argument in favor of something like democracy can arguably be traced back to the Sophists of ancient Greece and, specifically, to Protagoras (485–420 BC). While the Greeks themselves never provided a theory of democracy, according to the historian Finley, the Sophist Protagoras counts as "one exception, possibly the only one" (Finley [1973] 1985: 28). The particularly aggressive tone chosen by Plato in the *Protagoras* to voice Socrates's attacks against the eponymous character might well be explained by the fact that "Protagoras not only held characteristically Sophistical moral doctrines but also developed a democratic political theory" (Finley 1985: 28). Cynthia Farrar more boldly asserts that "Protagoras was, so far as we know, the first democratic political theorist in the history of the world" (1988: 77).[2]

In his famous Great Speech, Protagoras justifies the right of every citizen to speak in the assembly—*isegoria*—through what can arguably be redescribed as an argument from collective wisdom (i.e., collective intelligence, in my vocabulary). In what follows, I propose a reading of Protagoras's argument that probably differs from the meaning and intention of the historical character but, I believe, has some plausibility on its own.

Plato's dialogue pits Socrates, the Athenian philosopher, against Protagoras, the Sophist from Megara. Although the historical Socrates might have been a democrat, the character of Socrates in Plato's late dialogues tends to voice antidemocratic views. Plato's *Protagoras* thus reads to a large extent as an argument against democracy, opposing the figure of the real knower, who speaks the truth and bows to no one, to the figure of the crowd pleaser, who courts public opinion. In the passage of the dialogue that interests us, Socrates attacks Protagoras's pretension to teach the art of politics, remarking that both Athenian political practices and the way Athenians educate their children seem to rely on the assumption that there is no such teaching. As far as the Athenian political practices are concerned, Socrates thus observes that the assembly behaves very differently when the problem is to build an edifice or a ship than when it deliberates about the good of the city. In the first scenario, the assembly calls in architects and shipbuilders, and if someone who is not considered a competent technician in the relevant field speaks up to give his opinion, the crowd boos and shames him into silence. By contrast, Socrates goes on,

> When the question is an affair of state, then everybody is free to have a say—carpenter, tinker, cobbler, sailor, passenger; rich and poor, high and low—any one who likes gets up, and no one reproaches him, as in the former case, with not having learned, and having no teacher, and yet giving advice. (*Protagoras* 319d)

The reason why every man is entitled to a say in matters of state, Socrates concludes, is because people are under the right impression that "this sort of knowledge [about public affairs] cannot be taught" (ibid.). Socrates is trying to undermine Protagoras's pretension to teach politics by referring to democratic practices that rely on the assumption that political wisdom

---

[2] See Deneen (2005: chap. 4, esp. 120n7) for a list of the authors who consider Protagoras as a democratic thinker. Deneen also shows the roots of Plato's Protagoras and his "Great Speech" in Hesiod's *Works and Days* and *Theogony* and Aeschylus's *Prometheus Unbound*.

does not exist. Notice that here we can sense the tension between the real Socrates, who was endorsing democratic Athenian practices, and Plato, who in fact shares with the Sophist the belief that politics is a science but, contrary to the Sophist this time, believes that it is a science reserved to an elite few only. In this passage, Socrates expresses the Athenian democratic belief that everyone has a right to speak in the assembly but justifies it implicitly by the view that politics is not a form of knowledge (otherwise we would let the experts decide). Socrates further evidences his claim that there is no such thing as political knowledge with the fact that even Pericles, who otherwise had his children excellently instructed, failed to transmit to them his own political wisdom or have it imparted to them by someone else.

To the Socratic thesis that political wisdom cannot be taught because it is not a form of knowledge, Protagoras replies with a myth and an argument (*muthos* and *logos*). I will argue that the combination of this myth and argument implies that while politics is a form of knowledge, it is a form of knowledge that we can only attain collectively. It is because political knowledge is best known by the group rather than anyone within it—including experts and philosophers—that everyone has a right to speak, not because no one can know anything. Let me explain.

The myth depicts how man was originally left without any of the qualities necessary to survive by the careless Epimetheus, who forgot to distribute means of protection to human beings. After different failed attempts at solving the problem, Zeus charged Prometheus with correcting the situation by endowing man with reverence and justice, the two principles of cities, and the bonds of friendship and conciliation. In the myth Zeus insists that these virtues should be distributed equally to all, as opposed to only a few, as is the case for technical skills that only a few select people possess. Zeus orders a universal distribution of political wisdom—as reverence and justice combined—"for cities cannot exist, if a few only share in the virtues, as in the arts" (*Protagoras* 322d).[3]

The myth contains an implicit answer to Socrates's initial question about the difference between technical and political matters and who has legitimacy to speak up about them. Unlike the technical skills of carpentering or shipbuilding, which exist only in a few trained people, the virtue and art of politics are common to all men: "All . . . have a share" (ibid.). In other words, every man is born with enough political wisdom to have a legitimate right to speak about public matters. This myth, Protagoras

---

[3] Zeus adds: "And further, make a law by my order, that he who has no part in reverence and justice shall be put to death, for he is a plague of the state."

suggests, provides the rationale behind the Athenian practice of letting ordinary people speak up in the assembly.[4]

Protagoras's argument is that this innate ability for political wisdom—a gift from the god—can be cultivated and developed through education. Unlike what Socrates implies, political virtue is not something that one simply does or does not have. Everyone has it, but not everyone cultivates it. What Protagoras does not fully answer, though, is why this innate political wisdom translates into *isegoria*, or the equal right of speech in the assembly. Why is it that regarding governmental affairs, anyone, the most ordinary citizen and the most eminent—the *idiot* no less than the *aristos*—is qualified enough that his opinion should at least be heard?[5] After all, nothing in the myth says that everyone has exactly the same amount of wisdom, and since not everyone cultivates this virtue equally, one wonders why the inequality in background should not translate into an unequal say in political discussion.

One way to answer this question is by appealing to the phenomenon of collective wisdom. According to the myth, political wisdom is less a property of individuals per se—they could not necessarily do much on their own with their spark of political wisdom—than of the group as a whole. The usefulness of every human's fragment of political wisdom comes from the fact that it is distributed across all human beings and therefore most useful to individuals when they are gathered together in cities. The truth in political matters is more likely to be found by the group as a unit than by any individual within it, even if some individuals may develop their political wisdom as, arguably, the Sophists and their disciples do.

This interpretation of the myth in terms of distributed wisdom, however unconventional, helps us understand why the Sophist can at the same time assert that there is something to be taught and learned about politics, and yet respect the democratic principle of *isegoria*. Some may develop their ability for political wisdom more than others, but it is only by including everyone that the city will tap divine wisdom, in the form of collective wisdom—that is, the virtue initially distributed equally to human beings by the gods. To the extent that the myth may be interpreted

---

[4]The democratic potential of the argument contained in that myth goes far beyond the implications that Protagoras derives for the Athenian Assembly. The myth does not specify that only men, or only Athenians, or only Athenian free men, received a share of political wisdom. It says that all humans share in that god-given political wisdom. In theory, this should imply, if not universal suffrage, at least the equal right to speak in front of the assembly. Yet, in practice, only free men were allowed to speak up in front of the assembly. Aristophane's assembly of women, while it ridicules the very idea of women as politically involved, suggests that some democrats may have defended that idea.

[5]See, e.g., Manin 1997: 52.

as evoking the emergent phenomenon through which a group makes better decisions than any of its members, it is, however, hard to pin the myth as epistemic. Political wisdom is presented as an art, a virtue, but only ambiguously as a form of knowledge, let alone a science. As an argument for the collective distributed political intelligence of the people, the epistemic argument for the egalitarian and democratic principle of *isegoria* contained in the myth of Protagoras remains inchoate. One must turn to Aristotle to find it explicitly articulated.

2. ARISTOTLE'S FEAST: THE MORE, THE WISER

In the *Politics*, Aristotle famously compares democracy with a feast to which the many contribute. Jeremy Waldron (1995) has felicitously labeled this argument "the doctrine of the wisdom of the multitude" (or DWM, in his terminology). Aristotle introduces it in book III, chapter 11, of the *Politics* in the following terms:

> [T]he many, who are not as individuals excellent men, nevertheless can, when they have come together, be better than the few best people, not individually but collectively, just as feasts to which many contribute are better than feasts provided at one person's expense. (*Politics* III, 11, 1281a41–1281b2; trans. Reeve 1998: 83)

Waldron argues that there are two ways to interpret this argument, one stronger than the other. In the weaker argument, Aristotle may be interpreted as saying that "the people acting as a body are capable of making better decisions, by pooling their knowledge, experience, and insight, than any individual member of the body, however excellent, is capable of making on his own" (Waldron 1995: 564). Waldron calls this weaker version DWM1. A stronger version makes the case for democracy not only against kingship (the rule of the one best) but against aristocracy too, a case more difficult to make since, as Waldron observes, "an aristocratic regime may itself benefit from the doctrine." In fact, a group of very smart aristocrats may plausibly be as smart as a large group of average people. The stronger version, DWM2, argues that "the people acting as a body are capable of making better decisions, by pooling their knowledge, experience, and insight, than any subset of them acting as a body and pooling their knowledge, experience, and insight of the members of the subset" (p. 565). In other words, according to DWM2, the people as a group are smarter than all aristocratic subsets within the group.

This book generally supports the stronger epistemic claim in favor of democracy so it is tempting to read in Aristotle a precursor of that strong argument. The passages in which Aristotle defends the epistemic claim for democracy, however, are too terse to justify favoring one interpretation over the other. This is in part because it is not entirely clear what the meaning of the analogy of the feast really is or what kind of mechanism Aristotle had in mind to explain how an assembly of average people can turn into a smart group.

For most commentators (e.g., Waldron 1999 and Risse 2001), the mechanism that Aristotle has in mind is deliberation. On this reading, Aristotle gestures toward a deliberative model of party hosting. He can then be interpreted as saying that when several people pool their talents, ideas, and connections to throw a party, they are likely to be better at it than just one person. Moreover, a group is likely to bring a greater variety of food, wines, music, and invitees. A group of people will very probably give a better party than a single individual. In this deliberative model, the members of the group discuss and plan the way to organize the party. Perhaps Anaximandre will take charge of the music because he is a famed flute player in Athens, and Meno will command the wine because he is an enlightened oenologist. Or perhaps they will all make common decisions about everything.

On another interpretation, however, Aristotle could be suggesting something akin to an information-pooling mechanism that bypasses deliberation altogether (Manin 2005).[6] Another way to understand the analogy of the feast is thus statistical, rather than deliberative. On the statistical, aggregative reading, Aristotle had in mind a party in which guests contribute whatever they want—a cake, a bottle of wine, flowers, music, and so on. If the number of people is large enough, the chances that everyone will bring wine and no one bring music is very low. So the mere pooling of people's unconcerted contributions might end up producing a pleasant party, despite no one having planned the party. This type of spontaneous party might even be more fun than a party that a single person would have organized, especially if that person had idiosyncratic tastes (being, perhaps, allergic to flowers and only enjoying beer and heavy metal music).

Although the deliberative interpretation seems more plausible than the statistical interpretation, if only because the notion of probability on which a statistical account relies would not have been available before the middle of the eighteenth century, there is textual evidence for both readings. As a consequence, whether the epistemic properties of democracy lie in majority rule as opposed to some other mechanism like deliberation is not clearly evident in Aristotle's argument.

Let me now take a look at passages neglected by Waldron. A second version of the argument for the wisdom of the multitude crops up in

---

[6]Manin (2005: 17) remarks in particular that nowhere in the famous passage does Aristotle employ the notion of deliberation (*sumbouleuein*).

chapter 15 of the *Politics*, in the context of an argument about whether the rule of the one best man or the rule of all is superior (Aristotle temporarily disregards the intermediary case of the rule of the few best):

> Taken individually, any one of these people [members of the assembly] is perhaps inferior to the best person. But a city-state consists of many people, just like a feast to which many contribute, and is better than one that is a unity and simple. That is why a crowd can also judge many things better than any single individual. (*Politics* III, 15, 1286a27–33; trans. Reeve 1998: 94)

This passage helps us understand that the mechanism behind the wisdom of the multitude is probably deliberation (rather than sheer aggregation or summation of views). Aristotle roots his argument in the deliberative practice of Athenian "assemblies," which "come together to hear cases, deliberate and decide." (*Politics* III, 15, 1286a25–26; trans. Reeve 1998: 94)

The third version of this argument comes after the cynical argument that some element of democratic governance (or at least the appearance of it) is required in order to keep the masses satisfied and quiet. Aristotle feels the need to supplement this realpolitik claim by insisting that participation of the masses is not just a necessary evil, but also presents an epistemic advantage:

> For when they [the poor many] all come together their perception is adequate, and when mixed with their betters, they benefit their states, just as a mixture of roughage and pure-food concentrate is more useful than a little of the latter by itself. Taken individually, however, each of them is an imperfect judge. (*Politics* III, 11, 1282a34–38; trans. Reeve 1998: 84)

This passage raises again the idea that a group of people might be better than any of its parts. Here, however, Aristotle is more interested in justifying the role of the democratic element in the polity (the mixed regime he deems best) than defending democracy for its own sake. The democratic element is useful in combination with the aristocratic one, just as, if one may thus interpret the argument about the mixture of roughage and pure-food concentrate, bread is necessary to butter, or rice to meat. The cognitive abilities of the nobler element are actually enhanced by those of the more vulgar. This passage almost reads as an argument in favor of the educative property of political participation. When the many enter in contact with the fewer, smarter people, the cognitive capacities of the group—collective "perception"—may actually improve collectively to a level above the cognitive capacities of any individual within the group.

Aristotle advances a fourth and negative version of the epistemic argument in favor of democracy. Instead of saying that the many are better judges than one or a few individuals, he argues that they are less likely to judge poorly, because their judgment is less easily corrupted. Building yet again on the feast analogy, Aristotle supplements it with this additional clarification:

> Besides, a large quantity is more incorruptible, so the multitude, like a larger quantity of water, are more incorruptible than the few. The judgment of an individual is inevitably corrupted when he is overcome by anger or some other passion of this sort, whereas in the same situation it is a task to get all the citizens to become angry and make mistakes at the same time. (*Politics* III, 15, 1286a31–35; trans. Reeve 1998: 94)

The corruption mentioned here is cognitive, not moral. Unlike Rousseau, Aristotle is not worried about the possibility that citizens could become selfish and not vote for the common good. Rather, he worries about cognitive failures caused by passions such as anger. A group, suggests Aristotle, is less likely to be blinded by passion than a single individual or a small group within it. For a large group to have its judgment corrupted by passion, everyone in the group would have to become angry or jealous or envious, all making the mistake at the same time. This is, at least, the empirical assumption on which Aristotle seems to rely. Aristotle's argument may sound strange to us moderns, who are familiar with analyses of the "psychology of crowds" that demonstrate how, in fact, large groups are susceptible to the same passions as individuals. Yet, if we contrast the inertia of large groups to the emotional impulsivity of some individuals, the argument does have some plausibility.

Finally, a fifth, subtle version of the argument for the wisdom of the multitude can be found in chapter 16, after Aristotle has spent some time considering the difficulty that a single ruler has in overseeing many things at once, and the necessity for him to appoint numerous officials. At the end of the day, Aristotle wonders, what is the difference between the rule of a king, who has to delegate most decisions to many officials, and the more direct rule of those officials?

> Besides, as we said earlier, if it really is just for the excellent man to rule because he is better, well, two good ones are better than one. Hence the saying "When two go together . . . ," and Agamemnon's prayer, "May ten such counselors be mine." (*Politics* III, 16, 1287b10–14; trans. Reeve 1998: 98)

From two to ten to many, Aristotle seems inclined to follow the slippery slope that leads from the rule of one to democracy.

Later in the same paragraph, Aristotle also remarks that where the law is silent, "there should be many judges, not one only." He justifies this claim by arguing that

> [E]ach official judges well if he has been educated by the law. And it would perhaps be accounted strange if someone, when judging with one pair of eyes and one pair of ears, and acting with one pair of feet and hands, could see better than many people with many pairs, since as things stand, monarchs provide themselves with many eyes, ears, hands, and feet." (*Politics* III, 16, 1287b25–30; trans. Reeve 1998: 98)

Again, in this passage, Aristotle seems to slip from the idea that many experts are better judges to the idea that many—*simpliciter*—are better judges. The mere possession of ears, hands, and feet seems to substitute for accuracy of perception and judgment. Numbers, in other words, beyond a certain threshold, compensate for lack of individual accuracy.

In the end, Aristotle concludes that under certain circumstances, democracy *is* the most appropriate regime:

> [I]t is possible that the quality belongs to one of the parts of which a city-state is constituted, whereas the quantity belongs to another. For example, the low-born may be more numerous than the well-born or the poor more numerous than the rich, but yet one may not be as superior in quantity as it is inferior in quality. Hence these have to be judged in relation to one another. *Where the multitude of poor people is superior in the proportion mentioned* [so as to offset their inferior quality], *there it is natural for a democracy to exist.* (*Politics* IV, 12, 1296b16; trans. Reeve 1998: 122; my emphasis)

When the number of lowborn more than compensates for their inferior quality, democracy—and not the mixed regime Aristotle calls "polity"—is the best regime. Aristotle thus first exposed a clear argument for a direct form of democracy: the fact that many can be smarter than one or a few when it comes to deciding public issues. Aristotle does not provide us, however, with a complete and decisive argument as to how collective intelligence can emerge from the exchange of arguments and information among the many.

Nevertheless, Aristotle's ideas remained influential for centuries. Subsequent commentators such as Aquinas rehearsed Aristotle's arguments with little variation. Aquinas's pupil, Peter of Auvergne, used his commentary to oppose the idea of the multitude unfit to rule with that of the multitude where all have some share in reason and are therefore amenable to rational persuasion (see Bull 2005: 31n45). In the case of the reasonable multitude, Peter of Auvergne remarks, rule of the multitude is better than that of a few wise individuals. Marsilius of Padua also pressed home the point that "the common utility of a law is better known by the entire multitude."[7] We need to turn to Machiavelli, however, to see a new twist on the argument for the wisdom of the multitude. Machiavelli now likens the voice of the people to that of God.

3. MACHIAVELLI: VOX POPULI, VOX DEI

Machiavelli may perhaps seem like an odd figure to invoke in a chapter devoted to advocates of the epistemic properties of democracy. Isn't his most eminent work, *The Prince*, a manual for monarchs, ending with an exhortation to Lorenzo de Medici and the elite Medici family to unite Italy under their rule? How could the advocate of monarchy count also as an advocate of democracy? Yet one finds in Machiavelli's works— true, much more so in the *Discourses on Livy* than in *The Prince*—many instances of an overt appreciation for the superiority of the people's judgment over that of princes themselves.

Here I would like to focus on chapter 58 of the first part of the *Discourses*, which is explicitly titled "The Multitude Is Wiser and More Constant than a Prince," in which Machiavelli defends the superiority of the people's judgment in political matters.

In this famous passage, Machiavelli is answering the Roman historian Livy and other past authorities, according to whom there is nothing more vain and inconstant than the multitude and relying on the shifting and unreliable people is as good as building on mud. Against that argument, Machiavelli advances the following:

> [A]s to prudence and stability, I say that a people is more prudent, more stable, and *of better judgment* than a prince. Not without cause may the voice of a people be likened to that of God; for one sees a *universal opinion* produce marvelous effects in its forecasts, so that it appears to *foresee its ill and its good* by a hidden virtue. (Machiavelli 1996: 157–58; my emphasis)

Comparing the people with the prince, Machiavelli thus extols three things about the people: their prudence, their stability, and their *better judgment*. Notice that for Machiavelli, political judgment is a category distinct from, though partially dependent on, prudence and stability. Political judgment

---

[7]*Defensor Pacis* [1324] 1956, trans. Alan Gewirth (New York), I.I2.5; cited in Bull 2005: 26. Marsilius of Padua combines in this sentence the argument from collective intelligence and a distinct Aristotelian argument for democracy, by which citizens are entitled to a say in collective decisions because, even if they are unable to design policies, they are in the best position to judge their effects, the same way that the public is a better judge of a work of art than the artist or the user of a house a better judge of its flaws and qualities than its architect (*Politics* III 11). This is also called the "shoe-pinching argument" (as per Dewey 1954: 207).

consists in being able to achieve "prudence and stability" as an end. Even if the people *are* naturally prudent and stable, they need this extra dimension of political judgment to *make* prudent and stable decisions.

Rephrasing his argument, Machiavelli gives us a clearer idea of what he means by judgment: "As to judging things, if a people hears two orators who incline to different sides, when they are of equal virtue, very few times does one see it not take up the better opinion, and not persuaded of the truth that it hears" (1996: 158). Judgment is thus the ability of a people to decide in favor of the better opinion and to identify "the truth that it hears." It is this ability that explains the people's superiority over a prince: they simply have better judgment. Conversely, even if the people can make mistakes, a prince is more fallible still. In Machiavelli's words: "If it [a people] errs in mighty things or those that appear useful, as is said above, often a prince errs too in his own passions, which are many more than those of peoples" (ibid.).

Three arguments for the superiority of popular rule are contained in the passages quoted so far (all part of one paragraph). All three arguably have epistemic features. The first argument is the argument of "*Vox populi, vox dei*," which equates the voice of the people to that of God himself. On one reading, Machiavelli is simply asserting the willpower of the people, whose sheer number gives performative force to their collective will. Indeed, when many people happen to want the same thing, the resulting collective will might prove irresistible to even the most determined of tyrants.[8] Popular will could also be performative in the sense that it defines what the common interest is. The common interest may just be what the people, or at least a majority of them, want.

If one accepts that Machiavelli sincerely endorses the saying about popular wisdom, however, one way to interpret the saying "*Vox populi, vox dei*" is in epistemic terms, as the uncanny ability of the people to make judgments that track a procedure-independent standard of truth.

---

[8] A Straussian reading would thus emphasize that in its original context, after all, the extended version of that reference means the opposite of what it seems to mean here—namely that the people do not have the wisdom of God. The phrase "*Vox populi, vox dei*" is indeed usually attributed to the monk Alcuin, who advised Charlemagne that "those people should not be listened to who keep saying, 'The voice of the people [is] the voice of God,' since the riotousness of the crowd is always very close to madness" (*Works*, Letter 164, cited in *Little Oxford Dictionary of Quotations*, http://www.askoxford.com/quotations/2192?view=uk, accessed July 16, 2007). Machiavelli might thus in fact be reiterating the usual warning against the folly of crowds. On that interpretation, the voice of the people is performative, like the voice of God, because what the people believe will happen; they make it happen—just in the way there is light if God says there must be light. Machiavelli's argument for the superiority of the people's judgment over that of the prince would thus not be that the people are better predictors of the future but simply that the people make the future happen in accordance to what they think it will and/or should be. The argument here is that there is something self-fulfilling about the people's judgments about the future state of affairs.

Although Machiavelli leaves unanswered the question of how, exactly, popular beliefs turn into marvelous effects and accurate predictions, he does observe with what seems like genuine awe a phenomenon that we may identify as that of collective intelligence. Where Machiavelli saw a "hidden" and unexplained virtue, contemporary analyses unveil institutional mechanisms and psychological and social laws.

The second argument further supports an epistemic interpretation of Machiavelli's "*Vox populi, vox dei*" reference. Machiavelli argues that the people are more likely to make correct decisions than the prince. More precisely, given a choice between two opposite options, and after listening to the arguments on each side—provided the advocates of each position are of equal ability—the people will more often than not pick the "better" view and make up their mind correctly as to the "truth" of what they hear. This means that the people are capable not only of differentiating lies from truths in the orators' discourses but of identifying the better arguments. This ability to make good judgments is, in Machiavelli's account, passive and reactive. The people do not initiate a view; they simply respond to those voiced in the public forum.

In the same negative vein, the third argument stresses the lesser fallibility of the people. According to Machiavelli, the people are less likely to err than a prince because, as a group, they are less likely to be blinded by passions. Like Aristotle, Machiavelli does not explain what makes him so confident about the alleged immunity of groups to passions, relative to individuals. He does not explain whether this immunity is due to some group property (perhaps because many people's moods cancel each other out) or to a difference in virtue between the people and the prince (perhaps because the popular character is naturally less irascible than the princely one). In other passages, Machiavelli seems to suggest that the greater immunity to passion comes from the people's moral rightness and constancy when compared with the corruptibility and fickleness of the prince. Political competence, for Machiavelli, is an irreducible mix of the cognitive and the moral, of intelligence and virtue.[9]

Of course, as already noted, the explicit defense of the judgment of the people over that of the prince presented in chapter 58 of the *Discourses* does not square well with the more conservative and skeptical views expressed by Machiavelli in *The Prince*. However, in the conclusion of chapter 58 of the *Discourses*, Machiavelli gives us a hint as to what may explain this discrepancy. The reason why the prejudice against

---

[9] For a more thorough and excellent analysis of Machiavelli's defense of the superiority of popular judgment, see John McCormick's *Machiavellian Democracy* (2011), particularly chapter 3, "The Benefits and Limits of Popular Participation."

the masses is so widely shared is, he writes, "because everyone speaks ill of peoples without fear and freely, even while they reign; princes are always spoken of with a thousand fears and a thousand hesitations" (1996: 119). Earlier in the passage, Machiavelli had further noted that to remedy the faults of the people, "words are enough," whereas for the prince "steel is needed" (ibid.). In other words, princes are much more dangerous and less easy to criticize than the people. This explains not only the false belief in the superiority of princes but also gives us a hint as to where the true opinion of Machiavelli lies. While the *Discourses* were written for the people, whom Machiavelli could praise or criticize at little to no personal cost, *The Prince* was written for a member of the most powerful ruling family in Italy. Praise in one case is bound to mean much more than in the other.

4. Spinoza: The Rational Majority

Spinoza introduces both a continuation and a rupture in the linear story presented so far. He, too, was acquainted with Aristotle's argument and thus sounds familiar when he celebrates the desire for reason supposedly entrenched in the very nature of democracy. While Spinoza's *Tractatus Theologico-Politicus* was designed to keep the multitude at bay, as is clear from the recurrent motif of the masses as ignorant, prejudiced, and obstinate that he develops in the preface, it also contains a surprising celebration of the intelligence—in Spinoza's vocabulary, the "rationality"—of the people:

> In a democracy, irrational commands are still less to be feared [than in other forms of body politic]: *for it is almost impossible that the majority of a people, especially if it be a large one, should agree in an irrational design.* Moreover, the basis and aim of a democracy is to avoid the desires as irrational, and to bring men as far as possible under the control of reason, so that they may live in peace and harmony: if this basis be removed the whole fabric falls to ruin. (*Tractatus Theologico-Politicus* 16.16; my emphasis)[10]

[10] Given the slightly disquieting definition of a democracy for Spinoza—"a society which wields all its power as a whole" and in which "the sovereign power is not restrained by any laws" and "everyone is bound to obey it in all things" (*Tractatus Theologico-Politicus* [hereafter *Tractatus*] 16.14)—the argument comes as a relief. Spinoza's democratic inclinations—he insists that for him democracy is the most natural of all regimes, the best fitted to individual liberty, and the one that he acknowledges speaking of at greatest length (*Tractatus* 16.22, 23)—also contribute to making the argument sound sincere.

In this passage, Spinoza is saying that democratic majorities, especially large ones, are less likely to err than the ruling sovereigns of kingships and aristocracies. The explanatory mechanism for the rationality of democracy, however, is not strikingly Aristotelian. Spinoza makes no reference to deliberation but insists instead on the size of the majority: "for it is almost impossible that the majority of a people, *especially if it be a large one*, should agree in an irrational design."

It is tempting to see in this passage a proto-version of Condorcet's probabilistic and aggregative argument for majority rule based on the law of large numbers (to be explained shortly).[11] According to one commentator (Bull 2005), the epistemic properties of aggregation through majority rule make up the rationale that Spinoza advanced for expanding the numbers on a council. This is indeed the conclusion one may reach upon reading Spinoza's defense of large councils as having the most absolute authority ("dominion," in his language): "[T]he dominion conferred upon a large enough council is absolute, or approaches nearest to the absolute. For if there be an absolute dominion, it is, in fact, that which is held by an entire multitude" (*Tractatus* 16.2–5).

If Spinoza is truly concerned in the passage quoted above with the epistemic properties of judgment aggregation where large numbers are at stake, then his argument marks a real departure from the idea that collective reason is simply individual reason writ large. Indeed, the idea would no longer be, as in Aristotle and his followers, that the elements of reason present in each individual add up to something intelligent; rather the argument would be that the mistakes made by some will be overridden by the overall correct judgments made by others. On average, the group will choose rightly, even if individuals among them make mistakes.

While some elements of the passage suggest an aggregative interpretation, however, others invite a more traditional reading. Shortly after the passage where he insists on the size of the majority, Spinoza acknowledges the importance of deliberation in the collective decision-making process. Spinoza thus explains that

> men's natural abilities are too dull to see through everything at once; but by consulting, listening, and debating, they grow more acute, and while they are trying all means, they at last discover those which they want, which all approve, but no one would have thought of in the first place. (*Tractatus* 9.14)

---

[11]Whether the phrase "if it be a large one" refers to the size of the majority or the size of the people, both meanings can be reconciled with one version or another of the Condorcet Jury Theorem.

Malcolm Bull sees in this passage the confirmation that, for Spinoza, it is through the aggregation of individual judgments that reason prevails. Individual judgments might be wrong but collectively aggregated, through consultation and debate, they produce something correct or rational. In the end, it seems that for Spinoza, aggregation of judgments does not preclude deliberative elements and in fact is effective only to the extent that it involves deliberative activities (such as "consulting, listening, and debating").

5. ROUSSEAU: THE GENERAL WILL IS ALWAYS RIGHT

Rousseau's theory of the general will lends itself equally to a purely procedural interpretation, according to which the general will is whatever the people want, and an epistemic reading, according to which the general will is an independently given reality for which majority rule is a proxy. In the following, I will emphasize the epistemic interpretation.

One of the most famous passages from *The Social Contract* indeed suggests an epistemic approach to politics:

> When a law is proposed in the People's assembly, what they are being asked is not exactly whether they approve the proposal or reject it, but whether it does or does not conform to the general will, which is theirs; everyone states his opinion about this by casting his ballot, and the tally of the votes yields the declaration of the general will. ([1762] 1997: 124)

In this passage, Rousseau can be read as saying that when voting, citizens do not simply express a preference as to the proposal (whether they like it or not) but rather express an "opinion"—that is, in eighteenth-century French, a judgment—as to whether or not the proposal is in agreement with an independently given standard of political rightness, also called the common interest or the common good. Rousseau further suggests that majoritarian outcomes should be equated with such political rightness, an equation that allows him to conclude in what follows this passage that "when the opinion contrary to my own prevails, it proves nothing more than that I made a *mistake* and that what I took to be the general will was not" (p. 124; my emphasis).

Other passages, such as the chapter titled "Whether the General Will Can Err," equally invite an epistemic reading (p. 59). In this chapter, Rousseau explains why the general will can never be mistaken, even though individual citizens themselves can:

> There is often a considerable difference between the will of all and the
> general will: the latter looks only to the common interest, the former
> looks to private interest, and is nothing but a sum of particular wills;
> but if, from these same wills, one takes away the pluses and the mi-
> nuses which cancel each other out, what is left as the sum of the differ-
> ences is the general will. (p. 60)

Something like the emergent phenomenon of collective intelligence can
be identified in this famous distinction between the will of all and the
general will. Rousseau's account of the general will reminds one of what
chapter 6 will analyze as the "Miracle of Aggregation," in which the right
answer emerges at the collective level from the canceling out of individuals' mistakes.

The general will, which I interpret here as the procedure-independent
standard of correctness for majoritarian decisions, is often criticized for
being too abstract and unintelligible a notion. Yet Rousseau gives us later,
in book 3 of *The Social Contract*, a clear example of what the empirical outcome of political decisions aligned with the general will would
look like. In his view, the sign of a healthy polity, one whose decisions
are good for its people, is population growth (p. 105). Thus one could
argue that while population growth is not the general will, it is a sign that
political decisions approximate it well enough. In other words, population
growth is evidence that political decisions have, overall, approximated the
procedure-independent standard of correctness, that is, the general will.

Commentators have often drawn a comparison between Rousseau's
epistemic conception of voting and the Marquis de Condorcet's famous
Jury Theorem, to which I shall now turn.

6. Condorcet: Large Numbers and Smart Majorities

Condorcet's historical position as the last of the classical probabilists
and the first of the nineteenth-century statisticians allowed him to formulate a truly new breed of argument for majority rule, known as the
Condorcet Jury Theorem (CJT). Building on Bernoulli's "law of large
numbers" in the recently emerged science of probability,[12] Condorcet
gave mathematical rigor to the argument of the wisdom of the multitude.

---

[12] I borrow the formula from the title of Ian Hacking's book *The Emergence of Probability* (1984). Condorcet finds himself at the historical juncture between classical probability
and the new statistical theories developed in the nineteenth century. He is, however, more
plausibly considered the last of the classical probabilists rather than the first of the modern
statisticians, the latter title being more adequately applied to his contemporary Laplace.

To some, Condorcet's method reads as "a statistical hypothesis test" (H. P. Young 1988: 1235).

Condorcet published the CJT in his 1785 *Essay on the Application of Mathematics to the Theory of Decision Making* but it was only extracted from that lengthy essay and properly understood in the 1950s by Duncan Black (1958).[13] The CJT demonstrates that among large electorates voting on some yes-or-no question, majoritarian outcomes are virtually certain of tracking the "truth," as long as three conditions hold: (1) voters are better than random at choosing true propositions; (2) they vote independently of each other; and (3) they vote sincerely as opposed to strategically. The epistemic property of majority rule—the fact that, at the limit, majorities are virtually certain to be right when some assumptions are verified—is no more than the general property of large numbers. If voters are more often right than wrong on some yes-or-no question, then the majority opinion will be almost sure to be correct as the number of voters tends toward infinity.

This property is intuitively obvious where there are only two options. To illustrate, consider the following case. Suppose there are ten voters, each of which has a .51 probability of being correct on any yes-or-no question. Suppose the majority obtained consists of six persons. Then, by a simple calculus of conditional probability, we can say that the probability that the majority will be right is .52.[14] Now, extend the pool of

---

[13]The denomination "jury theorem" is, in effect, somewhat misleading. The term "jury" was given by Black, who did not understand Condorcet's political philosophy and thus wrongly suggested that Condorcet's conception of social choice was limited to juries (Arnold Urken, personal communication). Further, Condorcet's result was not phrased as a "theorem" in the modern sense of the term, with logical sequences of statements explicitly and rigorously derived. Condorcet himself referred to his mathematical arguments as "hypotheses" (Urken 1991: 17n3).

[14]The calculus is an application of Bayes's rule. Let us assume that there are two states of the world S and H, such that the probability of S is q and the probability for H is $1-q$; and that each voter has p chance to identify the true state of the world, and $1-p$ chance to identify the wrong state of the world. Now, conditional of k out of n people saying the true state of the world, call it X, is S, what is the probability that the true state is indeed X?

$\Pr(k$ out of n voters say $X = S | X = S) = B(n,k)\, p^k\, (1-p)^{n-k}$, where $B(n,k) = n!/(k!(n-k)!)$ is the binomial coefficient. Applying Bayes's rule:

$\Pr(X = S | k$ out of n voters say $X = S)$
$= \Pr(X = S$ and k out of n voters say $X = S)/$
$[\Pr(X = S$ and k out of n voters say $X = S) + \Pr(X = H$ and k out of n voters say $X = S)] =$
$\Pr(k$ out of n voters say $X = S | X = S)\Pr(X = S)/$
$[\Pr(k$ out of n voters say $X = S | X = S)\Pr(X = S) + \Pr(k$ out of n voters say $X = S | X = H)$
$\Pr(H)] = B(n,k)\, p^k\, (1-p)^{n-k}\, q\, / \, [B(n,k)\, p^k\, (1-p)^{n-k}\, q + B(n,k)\, p\, n{-}k\, (1-p)^k\, (1-q)] =$
$1/[1 + ((1-p)/p)^{2k-n}\, ((1-q)/q)]$

Assuming q = .5 and p = .51, for $k = 6$ and $n = 10$:
$P(X = S/6$ out of 10 voters say $X = S) = 1/[1 + (0.49/0.51)(2*6{-}10)] = 0.519992 \sim 52$.

voters to one thousand individuals. Mechanically, the probability that the majority—501 individuals in that case—will be right rises to nearly .73. As one keeps increasing the size of the group, the probability that the majority is right goes up to 1.[15]

Condorcet demonstrated a series of related results about the reliability of a decision body, depending on its size, the competence of its members, and the number of alternatives under consideration. These results, known as the "jury theorems," have been the object of some attention in the past twenty years (e.g., Grofman and Owen 1986; Grofman, Owen, and Feld 1983; Urken and Traflet 1984; Goodin and List 2001). One theorem shows that increasing the size of the relative majority increases its chances of being right mechanically just as increasing the number of voters.[16] For example, for a group of ten citizens, the probability of a majority of six of being right is .52 but the probability of a majority of eight being right is nearly .56.

Increasing citizens' individual judgment accuracy also increases the "competence" of the majoritarian outcome. As an individual citizen's probability of being correct goes up, so does the probability that the majority will be correct. The law of large numbers turns a small individual improvement into a huge collective advantage. When each individual's probability of being right increases from .51 to .52—a minor increase in individual's cognitive competence—the probability of the same majority of six people being right as a group increases from to .52 to .54.

As the number of people taking part in the decision grows infinitely large, or as the judgment accuracy of the average voter increases to the point of perfection, the probability that the majoritarian outcome will be right becomes closer to 1.[17] This seemingly magical—in fact, statistical—result holds true only under a set of relatively demanding assumptions, which Condorcet articulates as follows:

[15]Same reasoning as before, this time with $k = 501$ and $n = 1000$:
Pr(X = S/600 out of 1000 voters say X = S) = 1/[1 + (0.49/0.51)(2*600–1000)]
= 0.9996~1
As the number of voters grows large, the value of $2k–n$—which is the only variable that changes for a given majority proportion—becomes bigger and bigger, which in turn increases exponentially the probability that the majority is right.

[16]This does not necessarily mean that the more consensual a vote—that is, the closer to unanimity—the more likely it is to be right. On that issue, see Estlund and Goodin 2004 and also Feddersen and Pesendorfer 1998, refuted by Margolis 2001.

[17]H. P. Young (1988) illustrates the more complicated cases of voters faced with more than two options. He shows in particular that even if in some situations, Condorcet's selection of the winner (a variant of majority rule applied to multiple cases) is in competition with Borda's rule, there are strong probabilistic and axiomatic reasons to trust it as a ranking device.

We shall first suppose assemblies composed of voters possessing equal soundness of mind and equal enlightenment. We shall suppose that none of the voters influences the votes of others and that all express their opinion in good faith. (1976: 42)

The first assumption—the "enlightenment assumption" as I will call it— is later specified by Condorcet as the requirement that "the probable truth of the vote of each voter is greater than ½" (1976: 48). This require- ment is necessary in order for the majoritarian outcome to approximate certainty as the number of voters increases. The majoritarian outcome will be correct only when voters are more likely to be right than wrong on some yes-or-no question (or are more often right). While Condorcet believed that *each* voter had to have the same better than .5 correctness probability, modernized versions of the theorem make this demand on the *median* voter only, allowing for a more plausible diversity along the spec- trum of cognitive abilities. Condorcet's assumption has also recently been formally extended to the case of a multiple-choice situation, in which the probability threshold can be lowered below .5.[18] In the following, I will stick to the simpler original case of a two-choice situation requiring a correctness probability greater than .5, as the problem of trusting the reasonability of the average voter is fundamentally the same regardless of the number of available options.

The second assumption is voters' statistical independence. This "inde- pendence assumption" requires that votes be like coin flips, that is, events without causal effects on each other. In other words, the probability of one person being right on any binary question should be the same regardless of the probability of another person being right for that same question.

The last assumption made by Condorcet is that voters vote their true mind about what they deem to be the right answer about the question at stake. For example, in the case of a jury, the intention of the jurors when they cast their vote must be to vote guilty if they think the defendant is guilty and vote innocent if they think she is innocent. This requirement rules out strategic or dishonest voting.

Condorcet devised his theorem in order to determine the optimal num- ber of jurors on a jury. Although it was born in a judicial context, the CJT has arguably a broader meaning, supporting the view that the collective wisdom expressed through majority rule potentially beats the wisdom of

---

[18] See the details in Estlund and Goodin 2004. In fact, according to H. P. Young (1988: 1232), it is generally acknowledged that Condorcet himself at least "showed how the same arguments [as the ones he resorts in his study of two-option choice] could be extended to the case of more than two decisions."

any single individual within the group, including so-called experts and wise men. Condorcet's Jury Theorem (and Condorcetian theorems in general, see Grofman, Owen, and Feld 1983: 261–78) thus provides an important piece of a defense of majoritarian decision making. Despite its promise, however, the theorem has failed to spark optimistic conclusions among theorists of democracy and has instead generated puzzlement and skepticism.

We will come back to the Condorcet Jury Theorem in chapter 6. Before I move on to the next author though, let me emphasize the originality of the Condorcet Jury Theorem by comparison with previous attempts at accounting for the usefulness of majority rule. Some commentators (Grofman and Feld 1988) have claimed a particularly striking parallel between Rousseau and Condorcet, interpreting in particular the Rousseauian requirement that citizens vote in the silence of their hearts as the equivalent of Condorcet's independence assumption. While the comparison is indeed tempting, there are at least three reasons why Rousseau can hardly be said to forecast Condorcet. First, the probabilistic framework of Condorcet's theorem does not fit Rousseau's theory. As Grofman and Feld grant, some of the basic probabilistic ideas needed to make sense of the Condorcet Jury Theorem were still largely unconceptualized in the 1750s, when Rousseau was writing *The Social Contract*. Second, it is not clear that Rousseau used mathematical concepts as more than rhetorical tools. Rousseau was trying to dress up his theory in the scientific clothes of his time (like Plato or Spinoza before him), without necessarily being fully successful at the task. Finally, and most importantly, I would argue that Rousseau's argument for majority rule is not truly epistemic (even though his *conception* of democracy is). Rousseau does not primarily justify majority rule because it is most likely to give us the right answer. Figuring out the right answer is not an epistemic problem for Rousseau, because he assumes throughout that the right answer in politics is always obvious (Manin 1987). Only when individuals prefer their private interest to the general good or when factions introduce the smokescreen of partisan interest is the will of the people at risk of erring. All that the citizens need to figure out the general will is a pure heart (i.e., a good, nonperverted nature) and a genuine desire to discover the general will (i.e., the common good) rather than promote their narrow self-interest. In Condorcet, the problem is strictly reversed. Condorcet takes it for granted that people have the right kind of will. The problem is that truth is not obvious, and it takes a majority with the right kind of cognitive abilities to discover it.

The difference between the Rousseauian concern that citizens be "pure of heart" and the Condorcetian enlightenment requirement is substantive enough that it leads the two authors to opposing political conclusions.

Because a good will cannot be delegated,[19] Rousseau prefers direct democracy to representative democracy (indeed, he loathes representative democracy). By contrast, Condorcet has no problem with representative democracy, which allows, in his view, a necessary and fair division of cognitive labor within large assemblies of people.

If Rousseau arguably counts as an epistemic democrat, he is definitely not a deliberative one. Rousseau does not believe in public deliberation and prefers that voters have "no communication among themselves" ([1762] 1997: 59). This is because although political truth is relatively self-evident for Rousseau, demagogues and rhetoricians can obfuscate that truth, misleading the most well-intentioned citizens. In that sense, too, he is very different from the Marquis de Condorcet, who never discarded the usefulness of deliberation.

Condorcet's Jury Theorem thus marks a truly groundbreaking way to account for the usefulness of majority rule. Unlike his predecessors, Condorcet describes precisely the probabilistic logic of majority rule as an aggregation mechanism and, as a result, allows for a clearer identification of the respective epistemic properties of majority rule versus those of deliberation.

7. JOHN STUART MILL: EPISTEMIC DEMOCRAT OR EPISTEMIC LIBERAL?

One can find in John Stuart Mill's philosophy elements of a proto-epistemic argument for democratic decision making for at least two reasons. First, Mill famously advances an argument in favor of what he calls "representative government," which is, in an essential and surprising way, an argument for the epistemic properties of government by discussion and popular participation. The fact that Mill ultimately judges direct democracy to be impractical and substitutes for it a more elitist representative scheme—itself not universally applicable to all nations, but only to the more enlightened ones—should not detract from the fact that in a straightforward comparison between the epistemic properties of rule by one and rule by the many, he comes out in favor of democracy. Moreover, where only representative government of a more elitist kind is feasible, he promotes crucial deliberative institutions, such as the famous "Congress of Opinions" meant to represent all the voices and opinions of the people in ways evocative of the contemporary ideal of deliberative democracy.

Second, and more famously still, Mill's general defense of a free and diverse society is based on the idea that the free exchange of ideas is conducive to both the emergence of new truths and the preservation

---

[19] "Power can well be transferred, but not will" (Rousseau [1762] 1997: 57).

of old ones. Assuming that a free and open society is, in fact, essential to the functioning of democracy, the argument of *On Liberty* for a free market of ideas presumably qualifies as an epistemic argument for democracy itself.

I say "presumably" because, ultimately, there are some tensions within Mill's philosophy that make him, at best, an epistemic democrat with a strong elitist twist.[20] Even if one accepts the most recent interpretation of Mill as a proponent of a conception of representative government based on "the judgment of the community" (Urbinati 2002), rather than as a strict elitist defending the rule of the competent few, Mill is probably more of an epistemic liberal—acknowledging the epistemic properties of a free exchange of ideas among the people outside the formal structures of government—than an epistemic democrat per se. Even so, there are some irreducibly democratic elements to his philosophy. He thus stresses that a government will benefit from listening to the debates of an assembly representative of all voices, including those of nonelites and nonexperts. The quality of the debate in civil society further depends for him on the very nature and form of the government, and vice versa. Only under a government that fosters an educated and active citizenry can free public opinion lead to the discovery of new truths and the vivification of old ones. Conversely, an educated and active citizenry will, in part, determine the quality of government. Thus, despite the fact that Mill did not necessarily think that everyone's vote should have exactly the same value as everyone else's, and even though he is said to have quoted with approval the remark "Some are wise, and some are otherwise,"[21] his elitist views make room for surprisingly radical democratic conclusions.

The first explicitly epistemic argument developed by Mill in favor of democratic institutions is to be found in the third chapter of his

---

[20] Some commentators argue that Mill was not in the strict sense a democrat, because of his views about plural voting: "A consistent viewpoint unites Mill's political thought from start to finish; but it is not, in the strict sense he would himself have adopted, the viewpoint of a democrat" (Burns 1957: 294). Other commentators argue forcefully that Mill was a consistent democrat. According to Robson, for example: "So when the question is asked, by whom should government be selected? There is no doubt about Mill's answer. He is a democrat. From the time of his earliest sympathy with the Roman populace to his last writings the theme of popular control runs through his thought" (Robson 1968: 224). See *Cambridge Companion to Mill* (Skorupski 1998) for more details about this controversy. Nadia Urbinati has interpreted Mill's views in a democratic direction. She thus argues that "[a]lthough he did not elaborate a theory of democracy, he neither restricted political liberty to the narrow sphere of state institutions nor made politics the business of the competent few. Rather, he sought to devise avenues of participation that could absorb the transformation of politics engendered by representation" (Urbinati 2002: 2).

[21] Reported in the *Cambridge Companion to Mill* (Skorupski 1998; CW XXIII: 497).

Considerations on Representative Government, where he advances at least three arguments in favor of what he calls representative government but, in effect, describes as the advantages of a participatory, almost direct, form of democracy. The chapter is supposedly about the merits of representative government, but is really concerned with the advantages of popular participation and direct rule or, at the very least, with the way representative government offers advantages comparable to those of direct rule.

The first two arguments are well known. One is the "educative" argument, according to which representative government contributes to forming the characters of the people as active rather than passive. The second argument, sometimes dubbed the "protective" argument, is that representative government offers a way to protect the interests of the many against those of the propertied few. Finally, one can identify a third, properly epistemic argument. According to Mill, the reason why it is better to have and cultivate the wisdom of the many rather than that of a single despot—even a most enlightened one—is because under despotism, "while there may be a select class of *savants*" and "a bureaucracy" that cultivate ideas with a regard to practical use, the rest of the citizenry is left at best with "*dilettante* knowledge," which make people diminish "in their intelligence" as well as their "moral capacities" (Mill [1861] 2010: 37–38; emphasis in original).

Mill suggests in this passage that deliberation among the many is the key to good government. The claim is somewhat awkwardly supported, in the negative, by the assertion that if you let an elite class make the decisions, you do not create an incentive for the people to take charge and cultivate in themselves the knowledge necessary for good self-rule. What Mill seems to be saying is that given the right incentives and opportunities, the people are perfectly competent for self-rule. Their dilettante knowledge is potential political knowledge that needs to be put to use in order to be actualized and bear its fruit. Mill trusts that when the people are put in charge, they will be able to understand and manage their own interests. Mill's epistemic case for representative government thus relies on a transformative argument that connects representative government to an active citizenry. Representative government is good, not only because it transforms a passive into an active people, but also because this transformation activates the people's dormant capacity for self-rule.

Mill's argument for democracy as representative government is based on the epistemic properties of deliberation. Democracy is best conceptualized for Mill as government by discussion. Combining the epistemic argument with the transformative and protective arguments, Mill can conclude that:

> [T]he ideally best form of government is that in which the sovereignty, or supreme controlling power in the last resort, is vested *in the entire aggregate*; every citizen not only having a voice in the exercise of that ultimate sovereignty, but being, at least occasionally, called on to take an actual part in the government by the personal discharge of some public function, local or general. (Mill [1861] 2010: 41; my emphasis)

Mill has in mind, as ideal forms of active participation for citizens at the local level (in addition to voting), jury service and parish duties.

Mill thus believes in the epistemic superiority of the rule of the many over the rule of the lone genius. He does not believe in an epistemic trade-off between quantity and quality, at least not when the choice is between the many and the one. Given the complexity of modern polities, indeed, it is much more plausible to assume that the cognitive labor involved in ruling a country must be shared among many people rather than entrusted to just one person. However smart the lone ruler may be, no one can command the necessary amount of information and knowledge for such a task.

Addressing the hypothesis that a "good despot" is the best ruler at the beginning of the chapter, Mill remarked that the "realization [of good results] would in fact imply, not merely a good monarch, but an all-seeing one"; for, indeed, a good despot would need to be "at all times informed correctly, in considerable detail, of the conduct and working of every branch of administration, in every district of the country" as well as capable of giving "an effective share of attention and superintendence to all parts of this vast field"—a very implausible assumption, given that a king's days are no longer than a poor labourer's (Mill [1861] 2010: 36). Failing omniscience, a good despot would at least need to

> be capable of discerning and choosing out, from among the mass of his subjects, not only a large abundance of honest and able men, fit to conduct every branch of public administration under supervision and control, but also the small number of men of eminent virtues and talents who can be trusted not only to do without that supervision, but to exercise it themselves over others. (Mill 2010 [1861]: 37)

This second assumption is hardly less implausible than omniscience, being in effect a meta-omniscience: perfect knowledge of the knowers. Mill is skeptical that even if we could find in anyone the extraordinary "faculties and energies required for performing this task in any supportable manner," one could convince this person to consent to undertake the task (ibid.).

Short of such a god or a "man of superhuman mental activity" (ibid.), the best alternative is therefore to divide the cognitive labor among many people. In fact, in Mill's view, even if such a superhuman king could be found, the rule of the many would still be preferable because only in a representative government are the people encouraged to develop an active nature, which is for Mill a valuable end in itself.

Unlike with most previous authors, Mill's utilitarianism makes it relatively easy to specify what he has in mind when he says that government by discussion among the many is "better" than the government by one or a few. As a sophisticated utilitarian, he conceptualizes epistemic superiority as that which ensures both general prosperity and a higher form of human life. More specifically, the goodness of the government can be assessed with respect to two questions: "how far it promotes the good management of the affairs of society by means of the existing faculties, moral, intellectual, and active, of its various members" and "what is its effect in improving or deteriorating these faculties" (Mill [1861] 2010: 41). The superiority of representative government thus lies in that "it is both more favourable to present good government, and promotes better and higher form of national character, than any other polity whatsoever" (p. 42). In other words, the standard of a good political regime is both quantitative, whether it promotes the greatest material good of the greater number, and qualitative, what kind of character it breeds (within the framework of national identities—a given in Mill's times).

Opinions expressed with respect to the best means to that general end are conceptualized by Mill as judgments positioned on a continuum from falsity to truth. This is what allows him to say, in justifying the inclusion of the voice of the "working man" in the political process:

> I do not say that the working men's view of these questions is in general nearer to truth than the other; but it is sometimes quite as near; and in any case it ought to be respectfully listened to, instead of being, as it is, not merely turned away from, but ignored. (Mill [1861] 1993: 405)

Different people will find themselves located differently on the continuum from truth to falsehood. But instead of listening only to those we can assume are closer to the truth, we should include everyone, as any opinion may contain a relevant element of truth disregarded by others.

The second aspect of Mill's philosophy that possibly qualifies him as an epistemic democrat is his utilitarian defense of freedom of thought and expression in chapter 2 of *On Liberty*. Considering whether members of a majority are entitled to silence even one dissenter among them, Mill famously argues that the majority has no such right and that even if all mankind minus one were of one opinion, mankind would be no more justified in silencing that one person than this person would be justified in silencing mankind (Mill [1859] 1993: 85). Were he to defend the minority's right of expression for the sheer sake of the individuals' own interests, Mill would merely count as a liberal. Yet Mill defends the individual's right to free expression for reasons that have less to do with intrinsically valuable individual rights and more with the interests of the group, and even more specifically with the interests of the group in its search for truth. Writes Mill:

> [T]he peculiar evil of silencing the expression of an opinion is, that it is robbing the human race; posterity as well as the existing generation; those who dissent from the opinion, *still more than those who hold it*. If the opinion is right, they are deprived of the opportunity of exchanging error for truth: if wrong, they lose, what is almost as great a benefit, the clearer perception and livelier impression of truth, produced by its collision with error. (Ibid.; my emphasis)

For Mill, it is groups, more than individuals—those who dissent from the silenced opinion "more than those who hold it"—that are most harmed by policies limiting free speech. The main argument for the right of minorities and individuals to express themselves freely is thus utilitarian, instrumental, and more specifically epistemic: it benefits collective intelligence and the collective search for truth. In other words, it is for the sake of the many that free speech matters.

Of course, there are two serious limitations to this interpretation of Mill as an epistemic democrat advocating the virtues of direct, participatory deliberation. First, as has been noted with some uneasiness by many commentators, there are his views toward "backward" and "barbarian" nations, which do not deserve access to the more enlightened stage of representative government quite yet. Thus, in *On Liberty*, he famously notes that he leaves "out of consideration those backward states of society in which the race itself may be considered as in its nonage" (Mill [1859] 1993: 13). Mill's typical nineteenth-century belief in progress leads him to believe that there is a development path toward representative government, which, in the early stages, warrants and demands the forceful and undemocratic intervention of "a ruler full of the spirit of improvement," that is, a progressive despot. To be clear, "Despotism is a legitimate mode of government in dealing with barbarians, provided the end be their improvement, and the means justified by actually effecting that end" (pp. 13–14). Some interpretations (e.g., Sullivan 1998) have stressed that this is not just a lapse of a nineteenth-century employee of the British East India Company but that Mill was very worried about building as much epistemic expertise as possible into government, which led him to endorse despotism in some cases. If this is true, then Mill qualifies as an epistemic thinker in general, but as an epistemic *democrat* only in the case of what he considers to be advanced societies.

The second caveat that needs to be introduced regarding the reading of Mill as an epistemic democrat proposed here has to do with the fact that in chapter 4 of *Considerations on Representative Government*, Mill seems to take back some of the more democratic and participatory claims made in chapter 3. At least, the actual institutional scheme he lays out there does not seem fully compatible with his earlier philosophical conclusions. After abruptly ending chapter 3 with the claim that involving everyone in government would not be practical (Mill [1861] 2010: 51) and spending chapter 4 surveying the reasons why a people might not be fit for representative government, Mill goes on in chapter 5 to delineate an actual project of representative government in which the function of the democratically elected legislature is merely to vote up or down on legislation that a committee of experts write. Though the legislature can propose the topics of that legislation, they have no authority to actually write the law.

Despite these limitations in terms of what Mill judges a democratically elected assembly can legitimately do, he nonetheless concludes that the deliberative function, for which an assembly is better fit than any individual (Mill [1861] 2010: 65), is of utmost importance. He thus writes:

> I know not how a representative assembly can more usefully employ itself than in talk, when the subjects of talk is the great public interests of the country. . . . A place where every interest and shade of opinion in the country can have its cause even passionately pleaded . . . is in itself, if it answered no other purpose, one of the most important political institutions that can exist anywhere and one of the foremost benefits of free government. (p. 74)

The passage is worth noting not just for its emphasis on the centrality of "talk" at the level of free political institutions but also for the fact that it describes the deliberative assembly as a rainbow of various interests and opinions, or what he earlier called a "Committee of Grievances" and a "Congress of Opinions" (p. 73). He even argues that this assembly should be "a fair sample of every grade of intellect among the people" rather than "a selection of the greatest political minds in the country" (pp. 74–75). As we will see in the following chapters, one of the key components of a fruitful deliberation oriented toward problem solving is the cognitive diversity—the diversity of intellects in Mill's terms—that goes into it. Some aspects of Mill's defense of the representative assembly as a congress of opinions, as well as his defense of a more inclusive and participatory form of politics, which would include for example the voices of the traditionally unrepresented (like the workers) for the sake of their idiosyncratic perspective on a given problem, can be read as anticipating in intuitive ways much later insights in the social sciences.

82   •   Chapter Three

8. Dewey: Democracy and Social Intelligence

Dewey wrote about democracy after its historical reality had already shifted from parliamentary to party democracy, with the extension of the franchise and the rise of political parties and mass media. In *The Public and Its Problems* (1927), he analyzed the public's difficulties in finding and recognizing itself in the age of mass democracy and the machine. While sharing his contemporaries' worry about an apathetic and inarticulate public—which Walter Lippmann at the time had famously dubbed "the phantom public" (Lippmann [1927] 1993)—Dewey argued that this phenomenon was but a temporary "eclipse." The public had difficulty in identifying and distinguishing itself in an age that complicated the relationship between individual and collective actions and results, but Dewey did not see this problem as unsolvable.

He thus opposed the stance of his contemporaries who wanted to reduce as much of politics as possible to sheer administrative and managerial tasks; he refused to tread the path that would lead many toward elitist theories of democracy. To the contrary, Dewey took seriously the epistemic competence of the people and, at the same time, called for educative reforms meant to improve that competence. Some passages in the *Public and Its Problems* contain clear assertions regarding the epistemic properties of deliberation and the existence of "social intelligence."[22]

In the *Public and Its Problems*, Dewey attempts, among other things, to resist an antidemocratic current, according to which democracy was a mere transitional phase between the rule of landed proprietors and the rule of the captains of industry and government was a matter of an elite group of experts or knowers of some kind. To counter that claim, Dewey goes back to an argument for democracy that he finds in Tocqueville. He claims that the strongest argument in favor of democracy is that it involves "a consultation and a discussion which uncovers social needs and troubles," which "forces a recognition that there are common interests, even though the recognition of *what* they are is confused," and finally which "brings about some clarification of what they are" (Dewey [1927] 1954: 206–7). Dewey also suggests that even if the public is not necessarily the best problem solver, it is nevertheless the best judge of where the problem lies; hence the need to consult it (ibid.). The epistemic argument in Dewey can be rephrased as follows: democracy fosters a discussion that clarifies the nature of preexisting common interests and goals. On my interpretation, such preexisting common interests and

[22] In the following, I focus mostly on this book. In *Democracy and Education* (1916), however, one also finds a global definition of democracy as not only a set of political institutions but a way of life rooted in a certain type of education.

goals play the role of the independent standard of correctness in Dewey's epistemic theory. By contrast, experts might know how to fix problems, but without the information coming from the people, their knowledge is bound to be misapplied, because the interests of a secluded class are removed from the interests of the people and because private knowledge in social matters is not knowledge at all (p. 207).[23]

Dewey then explains why democratic decision procedures are valuable. The point of voting, he suggest, is not so much the act of putting a ballot in the ballot box but the public discussion that precedes this act: "Majority rule, just as majority rule, is as foolish as its critics charge it with being. But it never is *merely* majority rule" (ibid.). What makes it valuable are, among other things, "antecedent debates, modification of views to meet the opinions of minorities, and the relative satisfaction given the latter by the fact that it has had a chance and that next time it may be successful in becoming a majority" (pp. 207–8). Dewey thus suggests that the value of majority rule lies in the discussion that precedes it, which gives minorities an opportunity to voice their views and influence the final outcome. Far from supporting a populist view of democracy, according to which the majority is always right, Dewey acknowledges that "all valuable as well as new ideas begin with minorities," even "a minority of one," but he also stresses that "opportunity be given to that idea to spread and to become the possession of the multitude" (p. 20). Even if it is true that there are only a few people with actual knowledge of the relevant ideas, the value of their ideas can only come to fruition when they manage to convince the rest of their fellow citizens that they are right. In democracy, minorities can use the pre-voting debate to that end.[24]

Dewey can thus be described as an epistemic deliberative democrat. For him, solutions to political problems can be brought about only by a democratic debate in which those who know communicate their knowledge to the rest of their fellow citizens. This idea presupposes an average citizen sufficiently intelligent to be educated. It presupposes that truth tends overall to triumph over lies, misinformation, corruption, and sheer bad faith. As Dewey puts it, "the essential need . . . is the improvement of

---

[23] According to Dewey, political knowledge is best described as a form of opinion, mutable in nature (as conditions change so does the nature of the relevant knowledge) and highly public (knowledge must be communicated to deserve the title of knowledge) (Dewey [1927] 1954: 178).

[24] By contrast, it would be harmful to hand over all the decision power to the most knowledgeable minorities, because, Dewey argues, "no government by experts in which the masses do not have the chance to inform the experts as to their needs can be anything but an oligarchy managed in the interests of the few. And the enlightenment must proceed in ways which force the administrative specialists to take into account of the needs. The world has suffered more from leaders and authorities than from the masses" (Dewey [1927] 1954: 208).

the methods and conditions of debate, discussion, and persuasion. That is *the* problem of the public" (Dewey [1927] 1954: 208). For Dewey, the hope of a solution is not utopian. It depends essentially on the general public accessing and appropriating the conclusions of social inquiries led by experts and artists alike. Dewey does not raise the epistemic bar so high as to render democracy impracticable:

> It is not necessary that the many should have the knowledge and skill to carry on the needed investigations; what is required is that they have the ability to judge of the bearing of the knowledge supplied by others upon common concerns. It is easy to exaggerate the amount of intelligence and ability demanded to render such judgments fitted for their purpose. (Ibid.)

In this passage, Dewey might seem to lower the bar so much as to end up arguing for a relatively passive citizenry, one useful to judge rather than invent, to sanction rather than initiate. In the context of mass democracy—and along with it, high citizen apathy and growing skepticism toward democracy—the capacity to judge and sanction might have seemed already fairly demanding. Dewey concedes that "the data for good judgment are lacking," which makes it impossible to know "how apt for judgment of social policies the existing intelligence of the masses may be" (p. 210). Nevertheless, Dewey argues that this intelligence can be raised by appropriate educative measures. In his view, effective intelligence is not "an original, innate endowment" and depends heavily on social conditions (pp. 209–10).

Dewey also suggests that collective intelligence is not so much a function of individual intelligence (the existence of geniuses in the population) as of "embodied intelligence"—the knowledge contributed by many people, in however small amounts, crystallized and fixed for the future in public knowledge. To illustrate, embodied intelligence explains how "a mechanic can discourse of ohms and amperes as Sir Isaac Newton could in his day. Many a man who has tinkered with radios can judge of things which Faraday did not dream of" (p. 210). Stressing his point, Dewey adds that what truly matters for social intelligence is less individual intelligent contributions as "the difference made by different objects to think of and by different meanings in circulation" (ibid.). In order to raise collective intelligence, the key is not so much to raise individual intelligence as to improve "embodied intelligence," or the level upon which the intelligence of all operates. Indeed, "[t]he height of this level is much more important for judgment about public concerns than are differences in intelligence quotients" (p. 210–11).

This belief in the collective dimension of even individual intelligence— making individual intelligence conditional on a certain level of social intelligence—supports Dewey's philosophy of education as well as his preference for a form of direct, face-to-face deliberation. For him, a community, including a political community, "must always remain a matter of face-to-face intercourse." Thus, the pursuit of what he calls the Great Society—a society that nurtures social intelligence and full intercommunication between individuals—must develop on the basis of small-scale, local communities. Family and neighborhood provide the ur-model for that sort of community (p. 211). For Dewey, "democracy must begin at home, and its home is the neighborly community" (p. 213).

Another crucial feature of democratic deliberation for Dewey, which derives directly from this face-to-face dimension, is that deliberation must ultimately be oral rather than written. As he writes, "the winged words of conversation in immediate intercourse have a vital import lacking in the fixed and frozen words of written speech" (p. 218). Thus, while social inquiry and the dissemination of its results can take place in print, "their final actuality is accomplished in face-to-face relationships by means of direct give and take." Against the view that citizens should be mere spectators—mere readers of news—Dewey insists that they must physically engage in democratic debate, speaking with and listening to each other (ibid.). When democratic citizens engage in dialogue with each other in the context of a local community, Dewey argues, there is no limit to the expansion of "personal intellectual endowment" that can be brought by "the flow of social intelligence" (ibid.). Dewey concludes by invoking the poet's words: "We lie, as Emerson said, in the lap of an immense intelligence" (ibid.).

Dewey thus argues that the public is capable of social intelligence, which he defines as a capacity to communicate and create knowledge. One of the reasons for the eclipse of the public, as diagnosed by Dewey, is the incapacity of scattered individuals to find, identify, and even express themselves as a community united by a common purpose and sense of shared destiny. True democracy can only be realized, Dewey thinks, if and when the public develops the skills that make it able to take advantage of their right to political participation.

9. HAYEK: THE DISTRIBUTED KNOWLEDGE OF SOCIETY

Hayek's theory of how the dispersed and local knowledge of individuals aggregate through market mechanisms into accurate prices would seem to be an important part of the story of democratic reason. Upon closer examination, however, it is not. Unlike deliberation and majority rule, the two mechanisms covered in this book, markets do not produce decisions or offer a way of making collective decisions that can be appropriated by the public as such. The invisible hand that regulates the market of goods or information has certainly something to do with the law of large numbers and the "Miracle of Aggregation" (to be studied in a later chapter), but in the end, the market is not a political decision procedure. It thus does not offer an alternative to the rule of the many, at least not on a par with alternatives such as the rule of one or the rule of the few. Nonetheless, Hayek's fascinating insights are worth mentioning here, as they further develop important Millian intuitions about the preconditions for the emergence of truth in the public sphere, and also because they provide an argument for differentiating between the sphere of democratic reason and that of the market. This distinction overlaps with his famous contrast between two kinds of order.

One of Hayek's great contributions is indeed the distinction between two types of orders: those that can be achieved through intentional, rational planning and those that are "spontaneous," by which he means "self-organizing" rather than natural. The first type of order, he calls *taxis*; the other, *cosmos*. Based on this distinction, Hayek famously offers a new definition of order that encompasses both rational and spontaneous types. An order is thus "a state of affair in which a multiplicity of elements of various kinds are so related to each other that we may learn from our acquaintance with some spatial or temporal part of the whole to form correct expectations concerning the rest, or at least expectations which have a good chance of proving correct" (Hayek 1982: 36). Hayek was writing against contemporary conceptions that intended to replace the spontaneous order of the market with the purely rationalistic one of political decision making. He was thus essentially reacting against then-dominant socialistic views, which merged the political and economic order by having all decisions related to the latter—how much to produce of what good, for what price, and so on—made by the former. This confusion of orders was for Hayek both impractical and dangerous.

A rationalistic (socialist or otherwise) approach to economics was impractical, in Hayek's view, because it presupposed that the central agency in charge of implementing the politically designed economic order had more computing capacities than could possibly be gathered at one time, even in the best heads of the best governmental teams. The impossibility of performing the calculus of the right economic order—that is, setting the price of consumer goods, the level of workers' wages, the levels of production, and other economic parameters so as to create a well-functioning whole—is due to two things, for Hayek. One is the sheer enormity of the amount of information to be processed; the other is the fact that, even assuming full information at a given time, this order is constantly evolving in such unpredictable ways that human planning could never catch up with it. To put it another way, there is no economic order outside of the actual transactions of free individuals.[25] This double complexity is why, according to Hayek, it is better to let that infinitely complex calculus be performed at every instant by the unthinking mechanism of the market.[26]

Not only is man-made order unable to deal with the infinite complexities of human interactions, such that it risks impoverishing them by too much regulation, it is also easily threatened by manifestations of dissent, paving the way toward more and more control and the transformation of the government into an authoritarian and potentially totalitarian organization.

In light of the complexity and emerging nature of the economic order, and the dangers of trying to design it by intention, Hayek thought it best to keep the interactions of individuals in the economic sphere as free from rationalistic intervention as possible. Of course, the extent of the "possible" here is bound to remain a matter of controversy. One may agree with Hayek's view of the market as a spontaneous order (as I think one should) and yet disagree with the extent to which it could be corrected by rational decisions. The debate crucially depends on empirical claims about the imperfections of actual markets and how rational intervention—political regulations—can help fix them. But the original insight remains powerful and, I believe, true: in some domains of human life, including economic interactions, "truth" (or the socially optimal outcome) is best attained as the result of horizontal and unfettered individual actions and exchanges rather than as the result of a rational order imposed from the top down, even if this rationally order is constructed through democratic deliberation.

What Hayek brings to the table, therefore, is not so much an account of democratic reason but rather a crucial distinction between the epistemic properties of the market, which are both arational and apolitical, and those of political procedures. In a way, this helps us better understand, retrospectively so to speak, the proper status of the "market of ideas" that characterizes Mill's ideal of a free society. Such an ideal, as we saw, cannot pursue a rational consensus of the kind that can be pursued in deliberative settings. It can't even produce any determinate output, the

---

[25]This is a reading by James Buchanan, "Order Defined in the Process of Its Emergence," to counter the perception that economic order is simply the product of an omniscient designing mind. For Buchanan, Hayek's insight is that even assuming full information at any given time, the cosmos of the market (or any other cosmos) could not be intentionally designed because it exists only as the emergent phenomenon of certain types of individual relations. See note at http://oll.libertyfund.org/?option=com_content&task=view&id=163 &itemid=282.

[26]Every economic transaction between individuals at any given time contains relevant information that is immediately aggregated in market prices. See Hayek 1945: 519–30.

way an aggregation of judgments through majority rule, for example, can. The Millian market of ideas merely produces a diffuse and constantly moving "truth," which emerges in a dynamic form from the clash between conflicting points of views and is presumably only fully accessible to an impartial observer somewhat removed from the scene, such as a historian or a foreigner. This "truth" can perhaps be interpreted as a variety of the spontaneous order that, according to Hayek, emerges from a regular market when it is properly governed by the invisible hand that turns private greed into public good and aggregates a mass of dispersed and unknowable information into public and accurate prices. It is the same decentralization resulting from private exchanges between unmonitored individuals that account for the triumph of "true" ideas in Mill and the emergence of "accurate" prices (in the sense of prices reflecting all the existing information) in Hayek.

Yet a free market—whether of ideas or widgets and services—does not qualify as a proper political mechanism. As such, the Hayekian problematic, and to a degree the liberal Millian problematic, are orthogonal to the problem of democratic reason. Hayek may help us figure out the proper cognitive division of labor between the polity (i.e., the public qua citizens or their representatives) and the market (i.e., the public as consumers or market participants), but once this question is settled—in this book, I assume it is—he cannot help us in any way figure out who within the polity, of the many or the few, are best left in charge.

Looking back at this brief history of the epistemic argument for democracy, we see that its forerunners fall roughly into two camps, the "talkers" and the "counters"—or the advocates of democratic deliberation versus the advocates of nondeliberative judgment aggregation. A few of them, such as John Stuart Mill or Friedrich Hayek, develop arguments that do not fall neatly in the categories of either deliberation or aggregation of judgment through voting but, precisely for this reason, help us appreciate what a genuine epistemic argument for democracy, rather than the market, would look like. We can now turn from the history and genealogy of the epistemic argument for democracy toward an analysis of the mechanisms of democratic reason.

CHAPTER FOUR

First Mechanism of Democratic Reason: Inclusive Deliberation

THE PREVIOUS CHAPTER IDENTIFIED in the history of the epistemic argument for democracy two distinct mechanisms responsible for the production of collective intelligence: deliberation and majority rule. All in all, however, none of these previous authors, whether the "talkers" or the "counters," ever fully spelled out a systematic epistemic case for democracy. The following chapters now turn to this task, on the basis of some insights that can be gained from the previous historical survey and from recent social scientific advances in the understanding of the phenomenon of collective intelligence.

The main lesson that can be derived from a look back at the history of the epistemic case for democracy is that some kind of integration between the insights of the talkers and the counters is needed. In the following chapters, I thus look at deliberation and majority rule as complementary procedures in a larger democratic decision process, rather than as alternative or rival mechanisms. Deliberation by itself is, indeed, rarely conclusive. Conversely, to rehearse Dewey's insightful comment, "majority rule . . . never is merely majority rule," in effect being valuable to the extent that antecedent deliberation of some kind has taken place (Dewey [1927] 1954: 207). While neither of these procedures can be solely credited for the epistemic properties of democracy as a whole, they nevertheless need to be studied separately so as to identify their respective epistemic properties and the ways in which they can be fruitfully combined. Chapters 4 and 5 will thus focus first on deliberation, which generally comes prior to voting in actual practice. Chapters 6 and 7 focus on the epistemic properties of majority rule.

One crucial insight that was missing from, or at least crucially underspecified in, proto-epistemic arguments for democracy can be found in recent social scientific findings: namely, the importance of "cognitive diversity" for the emergence of collective intelligence (Hong and Page 2001, 2004, 2009, 2012; Page 2007). Cognitive diversity, to be defined more extensively later on, refers to the variety of mental tools that human beings use to solve problems or make predictions in the world. They include different perspectives, interpretations, heuristics, and predictive models. Hong and Page's findings establish the profoundly counterintuitive claim that cognitive diversity is generally as important as, and in some contexts more important than, individual ability for the emergence of the phenomenon of collective intelligence. My specific contribution in the following set of four chapters is to exploit the potential of these findings for the epistemic case for democracy by drawing a connection between the inclusiveness of the democratic procedure and the presence of cognitive diversity.

In this chapter, which focuses on deliberation, I develop an original case for why *inclusive* deliberation, that is deliberation involving all the members of a given group, has greater epistemic properties than less-inclusive deliberation, involving only a few members of the group. The argument proposed here takes for granted that deliberation about political issues is a particular case of collective problem solving and that Hong and Page's results apply nicely in this context. These results establish that, in a problem-solving context, cognitive diversity actually matters more to the production of smart collective solutions than individual ability does. Applied to deliberative assemblies, this means that the property we should want to maximize is cognitive diversity of the group rather than individual ability. I then argue that the more inclusive the deliberation process is, the more cognitively diverse the deliberating group is likely to be, on average. In contrast, the less inclusive the group, the more likely it is to be comprised of homogeneously thinking individuals. As a result, I conjecture that all things being equal otherwise, deliberation among more-inclusive groups is likely to produce better results than deliberation among less-inclusive groups, even if those less-inclusive groups include smarter people. In other words, it is epistemically better to have a larger group of average but cognitively diverse people than a smaller group of very smart but homogeneously thinking individuals.[1] Finally, I consider the implications of this argument that "more is smarter" for representative assemblies. Since all-inclusiveness is simply not feasible, and since representative democracy so far remains the only option for our mass societies, what is the next best thing in terms of maximizing the cognitive diversity of the actual decision-making group? In particular, what kind of selection mechanism for the choice of representatives would seem most appropriate to achieve that goal?

1. DELIBERATION: THE FORCE OF THE BETTER ARGUMENT

The first, most obvious, and perhaps oldest mechanism that makes democracy an epistemically reliable decision procedure is deliberation. Deliberation

---

[1] Let me emphasize that I will not address the issue of what are the criteria of membership in this chapter or elsewhere in the book. I take the idea of a *demos* as a given and delegate to others the task of theorizing what makes a group a democratic unit (whether it is the principle of "affected interests" or some other criterion).

has come to mean so many things in the contemporary literature on deliberative democracy that it is worth explaining what I mean by it.[2]

First, let me propose that there exist two species of deliberation: (i) simple deliberation, which takes place among individuals; and (ii) distributed deliberation, which takes place in a more depersonalized way at the level of society. Distributed deliberation is the more complex notion and can be defined as a decentralized and fragmented deliberative process distributed over different people or clusters of people (as embodied, for example, by a certain editorial line in newspapers or TV shows) as well as over time. In order to understand that kind of deliberation, however, we need a clear grasp of how simple deliberation functions. This is what this chapter will focus on.

Turning to simple deliberation, and following Manin (2005: 14), I take it to mean:

1. The action of deliberating, or weighing a thing in the mind; careful consideration with a view to decision.

2. The consideration and discussion of the reasons for and against a measure by a number of councilors (e.g., in a legislative assembly). (*Oxford English Dictionary*)

The first part of the definition refers to individual deliberation—what Robert Goodin calls "internal deliberation" (Goodin 2003). As the definition emphasizes, internal deliberation takes place "in the mind" and is performed "with a view to decision." The second part of the definition, the most relevant for our purposes, refers to collective deliberation by "a number of councilors," typically the members of a legislative assembly. Interestingly, the definition fails to mention that this kind of deliberation, like individual deliberation, has a view to an actual decision. What the definition emphasizes instead is the "discussion of the reasons for and against a measure."

In Aristotle, one finds a similar definition of deliberation as an exchange of arguments for or against something, with some ambiguity as to whether that exchange must result in some final decision or not (Aristotle, *Rhetoric*, I, 2). Crucially, Aristotle's definition emphasizes the reasoning aspect of the exchange of arguments. A similar emphasis can be noticed in most deliberative democrats' definitions, for example, in Joshua Cohen's definition of democratic deliberation as the "public use of arguments and *reasoning*" (Cohen 1989: 17–34; my emphasis) or Bernard Manin's (above). Manin is particularly explicit in connecting the centrality of argumentation with a specific mode of reasoning. He writes that "a

[2] The literature on deliberation since the "deliberative turn" of the late 1980s is massive. The following in no way pretends to offer a survey, synthesis, or even an exhaustive extraction of all the relevant arguments and pieces of information from this huge literature.

communication process qualifies as deliberation only if the participants employ arguments, that is propositions aiming to persuade members of the decision making body" (Manin 2005: 14). According to him, "we say that we deliberate, whether individually or collectively, when we engage in a *distinctive mode of mental activity, more specifically a distinctive mode of reasoning*" (ibid.; my emphasis). In this chapter and the next, I will consider reasoning and argumentation the core of deliberation.

Deliberation is thus, at the minimum, a form of communication relying on arguments, which may or may not have a decision as its end goal. Most deliberative democrats, however, define deliberation as a decision procedure (e.g., Cohen 1989: 72, Schauer 1999). Like them, I will assume that deliberation is a decision procedure in the sense that it must aim for an end point (even if it fails to achieve it). I thus do not follow Philippe Urfalino (2007, 2010), for whom deliberation does not qualify as a decision procedure per se; for Urfalino, it is more accurate to consider as the actual decision procedure the stopping rule that brings deliberation to a close, whether that stopping rule is unanimous consensus or some other rule.

Specifically, deliberative democrats tend to assume that the ideal goal of a deliberative procedure is a rational agreement or consensus on the better argument (Habermas 1996). This consensus is supposed to gather the unanimous approval of individuals, who all acknowledge the superiority of one outcome over the others.

In a much-quoted passage, Joshua Cohen, one of the first deliberative democrats, writes: "Outcomes are democratically legitimate if and only if they could be the object of free and reasoned agreement among equals" (Cohen 1989: 22). The meaning of "free and reasoned agreement among equals" in this passage is generally interpreted as a requirement of unanimity on a particular outcome (e.g., Bächtiger et al. 2010: 37; Dryzek and Niemeyer 2006: 635), by contrast in particular with aggregative approaches to democracy in which a mere majority can impose their choice, without further reasoning, on a dissenting minority. Similarly, in another famous statement considered to be a founding principle of deliberative democracy, Habermas defines the "discourse principle" as the principle according to which "[o]nly those norms can claim to be valid that meet (or could meet) with the approval of all affected in their capacity as participants in a practical discourse" (Habermas 1990a: 66). Here the requirement of explicit approval by all is made the stopping rule for a legitimate decision or choice. Specifically, deliberators are expected, in the ideal speech situation where there are no time and information constraints, to reach an uncoerced agreement on the "better argument."[3]

---

[3] Even though Habermas also acknowledges, particularly in later writings, the legitimacy of majority rule and bargaining (a strategic rather than communicative form of action), he firmly maintains the priority of rational discourse and consensus on both.

Some authors have described and commended alternative and less-demanding deliberative stopping rules such as "decision by interpretation" (Steiner and Dorff 1980)[4] or "apparent consensus" (Urfalino 2007, 2010, 2012), also labeled "decision by non-opposition" (Urfalino 2011).[5] The corresponding "consensus" is less perfect than the rational unanimity envisioned by deliberative democrats, but it is a real and reasonable alternative that has been empirically documented as accounting for up to 30 percent of actual group decisions (see Steiner and Dorff 1980). Though I will not use this refined terminology in this chapter, it should be clear that what I mean by deliberation is the combination of an exchange of arguments and the stopping rule that brings this exchange to a close, either through "rational consensus" or what Steiner and Dorff call "decision by interpretation" and Urfalino calls "decision by non-opposition." In any case, it should be noted that both the goals of rational unanimity and apparent consensus will occasionally fail to be met, in which case a decision procedure other than deliberation, such as voting, can be resorted to. My view is that voting does not necessarily take second place to deliberation but is the necessary complement of deliberative procedures, endowed with its own distinct epistemic properties. (Most deliberative democrats, however, tend to see the occasional necessity of voting as a sign that actual deliberation falls short of the ideal.)

Deliberation is not only opposed to voting, it is also opposed to bargaining (Elster 1986). In that view, deliberation is not supposed to involve, in particular, threats, promises, sophistry, or any form of "strategic" rather than "communicative" action. The better argument is supposed to triumph through what Habermas famously calls its "unforced force" (Habermas [1977] 1984). This unforced force describes the kind of nonviolent persuasion that truth or a superior argument has on the mind of individuals who yield to it without physical force or bribery being used. As I understand it, this unforced force corresponds, minimally, to the persuasiveness of the most rational or reasonable position and, maximally, to the actual epistemic validity of an argument.

I have exposed so far what some have termed the "classical" definition of deliberation (Mansbridge et al. 2010).[6] Critics have argued against

---

[4]Steiner and Dorff actually distinguish three types of this decision mode: (a) the chair takes the sense of the meeting and nobody objects, (b) the sense of the meeting is revealed in the minutes and in the next meeting nobody objects, (c) the discussion takes such a turn that a proposal falls by the wayside, for example, with the discussion turning to questions of implementation of another proposal.

[5]Decision by nonopposition is the situation in which even people who are not sure that they fully agree with a given proposal decide to abstain from voicing an opposition to it—either for reasons of epistemic uncertainty, indifference, or deference to what they perceive as the majority's will or other reasons.

[6]I summarize here a huge literature on the subject, following a recent survey by Martí 2006.

this classical model on the basis that it fails to take seriously the fact that politics is ultimately about "power and interests" (e.g., liberal pluralists like Ian Shapiro 1999), that it illegitimately excludes nonrational, more emotional modes of expression, such as rhetoric, greeting, or storytelling (e.g., Sanders 1997; Young 2000) or that it fails to grasp the profoundly "agonistic" nature of the public sphere, in which "different hegemonic political projects can be confronted" (Mouffe 2005: 3).

For recent commentators, we now ought to distinguish two "camps": on the one hand, type I deliberation theorists, who stick to the Habermasian rational consensus as a valuable ideal, although they do so at some normative and empirical costs; and on the other hand, type II deliberation theorists who have engaged in a more empirical program rid of rational consensus—a program that has, however, several blind spots of its own (Bächtiger et al. 2010). What is certain is that many deliberative theorists have taken their distance from the ideal of consensus, particularly rational consensus, and come to reassess the value of disagreement. From a necessary evil, disagreement is now touted as an often desirable good (e.g., Gutmann and Thompson 1996; Thompson 2008: 508). Theorists have come to endorse the full legitimacy of stopping rules for deliberation that used to be considered "second best" forms of nonconsensus, such as majority rule or other forms of judgment aggregation, or even the kind of noncommunicative and thus non-"rational" (in a Habermasian sense) agreement reached through bargaining. These type II deliberative democrats, led by Jane Mansbridge, argue that "when interests and values conflict irreconcilably, deliberation ideally ends not in consensus but in a clarification of conflict and structuring of disagreement, which sets the stage for a decision by non-deliberative methods, such as aggregation or negotiation among cooperative organisms" (Mansbridge et al. 2010: 68).[7] In this approach, the ideal termination of deliberation is not agreement but disagreement, followed by a nondeliberative decision rule.

In what follows, I focus on the theoretically purer version of deliberation offered by classical deliberative democrats (e.g., Cohen 1989; Elster 1989a; Habermas 1996; Manin 2005) and thus embrace a type I deliberation approach to democratic deliberation. I do so because this classical version strikes me as a much cleaner ideal type to work with. The account of deliberation offered by type I theorists will allow us to gauge the epistemic properties of deliberation without risking its entanglement with forms of aggregation, which will be considered later.

---

[7]Mansbridge et al. (2010) not only give up on rational consensus as the normative ideal of deliberation but reintroduce in deliberation interests, power, and forms of strategic action such as bargaining.

Furthermore, contrary to what its critics suggest, using this conception of deliberation does not necessarily prevent us from taking seriously the role of emotions, rhetoric, or interests in actual deliberative exchanges. I would thus argue that rhetoric, storytelling, and emotions can be interpreted within the paradigm of classical deliberation as vehicles for rational arguments. In the example I develop later in this chapter, taken from the jury members in the film *Twelve Angry Men*, the jurors do appeal to emotional personal experiences—hence the adjective "angry" rather than "impartial" or "rational" in the title—sometimes phrasing their points in the form of rhetorical questions or eloquent parables. Yet, this storytelling and rhetoric ultimately boil down to a series of arguments that can be rationally expressed and evaluated. I think what Habermas and others have in mind when setting the requirements of an ideal deliberation is not a sanitization of deliberation from human emotions and "nonrational" means of communication but simply a consideration of what in these various and diverse forms of communication could be rephrased as rational outputs.

Thus, in response to the objection of agonistic pluralists for whom classical deliberation artificially pacifies politics and imposes a hegemonic norm of rationality, I must confess that I fail to see what other norm could legitimately take the place of rationality. In fact, while agonists tend to argue that every norm is just a hegemonic articulation of particular power dynamics, they themselves sneak norms into their own accounts—norms that sometimes very much resemble those of deliberative democrats (such as the ethos of respecting our debating adversaries as equals, whose opinions deserve to be heard). Moreover, regarding some type II deliberation theorists' claim that power and interests must be reintroduced in the concept of deliberation (e.g., Mansbridge et al. 2010), I agree, like most classical democrats would, only to the extent that these elements are useful pieces of information. However, I reject the view that raw interests (e.g., subsidies for farmers) have any justificatory value as such and should be placed on the same level with reasons and arguments in the deliberative process. Only interests that have been shown to be generalizable (e.g., it is in everyone's interest to give subsidies to farmers) start having some form of justificatory value. At this point, however, such generalized interests are embedded within an argument about the general good, and it is not the fact that they are in someone's interest that justifies them.

Finally, in my view, another reason to stick with the classical version of deliberation is that it has clear roots in, or at least minimally departs from, the notions of deliberation used by philosophers surveyed in chapter 3. We saw in that chapter that many proto-epistemic arguments for democracy identified in the tradition of political philosophy impute to deliberation among citizens the production of good public decisions and policies. This argument considers democracy not so much as rule of the many as government by discussion among the many. From Aristotle to Mill to Dewey to contemporary deliberative democrats, deliberation is indeed largely endorsed as a key democratic practice.[8]

We have now clarified the concept of deliberation as it will be used in this chapter. There remains, however, a certain ambiguity in the connection between deliberation and democracy. The question now worth asking is: What makes collective deliberation of the kind defined above specifically "democratic"? Deliberative democrats generally agree that deliberation is democratic to the extent that all participants are free and equal, all have an equal opportunity to influence the outcome, and the exchange of arguments is governed by the principle of reciprocity. This definition, however, hardly gives us the criterion of who ought to be included among the participants of deliberation in the first place. It is simply assumed that democratic deliberation is inclusive. What matters most is that deliberation must recognize in each individual an equal capacity for influencing the final decision (e.g., Cohen 1989).

From the perspective of the argument laid out in this chapter, the difficulty is to connect the epistemic properties of deliberation in general and those of *inclusive* deliberation. One could argue that deliberation has epistemic properties whether or not it is inclusive in the way assumed by deliberative democrats.[9] For example, as was raised in the preceding chapter, it might be better, in a large group, to let the smarter people speak and make decisions on behalf of everyone else. Indeed, past a certain numerical threshold, deliberation turns into a chaotic mess; would it not be more effective if it were limited to a smaller number of people, preferably the smarter or more educated ones—that is, an oligarchy of the wisest or, as David Estlund calls it, an epistocracy? What we need, therefore, is an account of the following: first, provided that deliberation

---

[8]Sometimes, however, deliberative democrats slip into a defense of the epistemic properties of deliberation that is distinct from a defense of democracy. After all, discussion can be conducted between two persons only. Deliberation can also occur within the self. Rawls's scenario of the original position thus hardly counts as an example of deliberative democracy, since the discussion that is supposed to take place among the participants could be conducted by a single person alone.

[9]This is probably why those epistemic democrats who rely on deliberation as the primary epistemic mechanism (e.g., Estlund 2008) do not assert the epistemic superiority of democracy over other regimes, since deliberation is something that can take place among fewer people as well—in an oligarchy or even in a dictatorship, to the extent that deliberation can take place inside the head of a single individual. Thus, David Estlund asserts only that the epistemic properties of deliberation ensure that democracy is smarter than a random procedure. Habermas (2006) does not venture into comparisons at all.

among the many is feasible, even in an indirect or "represented" way, we need to know why it has any epistemic properties at all; and second, in light of this first claim, we need to be able to defend on epistemic grounds the claim that inclusive deliberation has higher epistemic properties than noninclusive deliberation, that is deliberation among the few.

Before I can address the problem of why inclusive deliberation is actually superior to less inclusive deliberation, let us substantiate the claim that deliberation has epistemic properties in the first place.

2. DELIBERATION AS PROBLEM SOLVING: WHY MORE COGNITIVE DIVERSITY IS SMARTER

Let me for now bracket the question of the empirical success (or lack thereof) of deliberation—a question that I will document at greater length in the next chapter—in favor of discussing its theoretical epistemic qualities. Three main arguments have been classically advanced for the known epistemic properties of deliberation. Deliberation is supposed to

1. Enlarge the pools of ideas and information

2. Weed out the good arguments from the bad

3. Lead to a consensus on the "better" or more "reasonable" solution

These, however, are very general claims that do not really account for why we should expect deliberation to have those results. In the following, I will use a more fine-grained analysis, applying the insights of Lu Hong and Scott Page about the logic of collective intelligence in problem solving to the nature of deliberation between two people or more. Roughly speaking, Hong and Page show that when it comes to problem solving, cognitive diversity actually trumps individual ability. I take it that problem solving aptly describes a central activity of democratic assemblies, whether they are trying to fix the national health care system, come up with a way to regulate bankers' compensation levels, or find environment-friendly ways to develop the economy. In the following, I use the importance of cognitive diversity in problem solving as an argument for why deliberation among the many offers an edge over deliberation among the few.

Let me now illustrate how deliberation yields the effects listed above by decomposing the logic of deliberating groups in three stylized examples. I borrow the first example from the film *Twelve Angry Men*. The second is a stylized example of a deliberation among three French congressmen, or *députés*. Finally, the third one is a real-life example of local problem solving in a neighborhood of New Haven, Connecticut. These are all examples of direct, rather than representative, deliberative democracy. They are doubly convenient, first, because they offer a certain deliberative purity that often gets lost in national legislatures, as interests, partisanship and ideological posturing obscure the purely argumentative content of many debates. Second, the procedure-independent standard of correctness in each case is particularly obvious and intuitive. Ideally, however, the logic of deliberation in such settings should translate to deliberation among representatives and to complex questions in which the standard of correctness is much harder to make sense of.

In "Twelve Angry Men," the famous film by Sidney Lumet, one brave dissenting jury member—juror number 8, played by Henry Fonda—manages to persuade the other 11 jurors to reconsider the guilty sentence they are about to pass on a young man charged with murder. Asking the other jurors to "talk it out" before making up their mind, juror number 8 takes the group on a long deliberative journey, which ultimately ends in unanimous acquittal. Whereas "Twelve Angry Men" has largely been interpreted as a movie about the importance of courageous, lone dissenters braving mistaken majorities, I take it to illustrate, rather, the phenomenon of collective intelligence emerging from inclusive deliberation. The brave dissenting juror, juror number 8, would not have been able to accomplish much left to his own devices. In particular he would have been unable to substantiate his initial hunch and demonstrate that the sentence was beyond reasonable doubt. Only by harnessing the intelligence and cognitive diversity of the other members, often against their own passions and prejudice, does the group ultimately reach the truth.

The contributions of each jury member vary and complement each other: juror number 5, a young man from a violent slum, is the one who notices that the suspect could not possibly have stabbed his victim with a switchblade. No other juror was acquainted with the proper way to use a switchblade. Juror number 9, who is an old man, then questions the plausibility of the time it took one of the key witnesses to cross the corridor. One of the most rational jurors, a stock broker who is not convinced by any of the other arguments, finally has to admit that a nearsighted woman is not credible when she pretends to have seen the murderer from her apartment across the street, through the windows of a passing subway, while she was lying in bed, most likely without her glasses. The deliberation process in this scenario nicely idealizes real-life deliberative processes in which participants contribute an argument, an idea, or a piece of information and the group can reach a conclusion that no individual by himself could have reached.

Notice that in this scenario, deliberation among several people has the three properties of good deliberation. Deliberation enlarged the pool of information and ideas for all jurors, bringing to the surface knowledge about the proper use of a switchblade and a contradiction between this proper use and the description by the visual witness of the way the victim was supposedly stabbed. Deliberation also brought to the surface a fact that many in the group had noticed—the red marks on the sides of the nose of the witness—but did not know how to interpret or use. Deliberation provided the proper interpretation of that fact: that the witness wears glasses, is most likely nearsighted, and therefore that her testimony cannot be trusted.

Deliberation also allowed the group to weed out the good arguments from the bad. Once they reach the conclusion that the visual witness is nearsighted, the jurors ask themselves whether she was likely to be wearing her glasses while lying in bed. Even the most stubborn juror has to admit that the argument that she was not wearing her glasses is stronger than the argument that she was.

Finally, deliberation in this example leads to a unanimous consensus on the "better" answer, namely the decision to consider the young convict "not guilty" given the doubts raised by deliberation.

Now let us turn to an even more stylized situation, which should bring out the logic of collective intelligence in deliberation even more clearly.[10] Imagine that the French government is choosing a city for an experiment with a new program. Three *députés* are deliberating, one from Calvados, one from Pas de Calais, one from Corrèze. They are aware of different possible solutions (the cities between parentheses below), each of which have a different objective value for the experiment. On a scale from 0 to 10, a city with value 10 has the highest objective value for the experiment. Each of the cities that a given *député* might offer counts as a "local optimum" (a technical term referring to the solutions to the problem that each member arrives at after duly thinking about it). The goal is for the group to find the global optimum, that is, the city with the highest objective value.

> Calvados: Marseille (7), Caen (10)
> Corrèze: Paris (8), Grenoble (9), Caen (10)
> Pas de Calais: Grenoble (9), Caen (10)

Let us further assume that each *député* has a higher probability of getting stuck at his lowest optimum than at his highest one. Thus, even though Caen is the better choice, the *député* from Calvados is not likely to think of it first, because he thinks that only big cities like Marseille will work, or perhaps because he is subconsciously prevented from choosing the capital of his own *département*. Similarly, suppose that the *député* from

---

[10] I borrow and suitably modify an example from Scott Page (personal communication).

Corrèze is pushing Paris, which has a value of 8, over his other two local optima, Grenoble and Caen, and that the *député* from Pas de Calais is pushing Grenoble over his other optimum, Caen. For whatever reasons, none of the *députés* thinks of his highest optimum first. Here is where deliberation can help.

The *député* from Calvados might start by saying: "This program should be implemented in a big city, so I propose Marseille [7]." The *député* from Corrèze says: "Good idea, but then Paris [8] is better." The *député* from Calvados has to agree (the force of the better argument obliges). Then the *député* from Pas de Calais interjects: "Actually Paris is really expensive for the project, we would be better off applying it in a moderate-sized city, which will be just as good a test bed. How about Grenoble [9]?" The *député* from Corrèze agrees, but the *député* from Calvados then says: "Fine, but as far as moderate-sized cities go, Caen [10] is even better than Grenoble [9], and less polluted too." In the end, they can only end up at Caen.

Here again, deliberation among several people has the three properties of good deliberation. The pool of information was enlarged, as the *député* from Calvados, who only knew about two local peaks (Marseille and Caen), ends up knowing about the qualities of Paris and Grenoble as well. The *député* from Corrèze learns about one other local peak (Marseille) and the *député* from Pas de Calais about two others (Marseille and Paris). Notice that even if the information gained is sometimes of lesser objective quality than that which the person already held, nonetheless, only by acquiring it can the members of the group reach the highest local optimum with certainty. The *député* of Calvados might never have considered an option he knew about, Caen (10), if he had not been spurred away from his initial choice (of value 7) by the other two *députés* who offered other suboptimal solutions (of respective values 8 and 9).

Deliberation also allowed the group to weed out the good arguments from the bad. While it seemed at first a good argument to look for a big city (Marseille, Paris), it turns out that it was better to look into moderate-sized cities (Grenoble, Caen).

Finally, deliberation did lead to a consensus on the "best" solution, namely the solution that allowed the group to reach the optimum of 10, when the pre-deliberative beliefs about the best solution could have been respectively 7, 8, and 9.

Let me now turn to a third example, one that is more clearly akin to political problem solving than the jury example and, at the same time, more textured than the abstract model just presented.

The issue in a New Haven neighborhood, called Wooster Square, was the recurrence of muggings on the Court Street Bridge, which separates Wooster Square from the downtown area. In a first attempt at addressing the mugging problem, the neighbors organized block watches and started an online site allowing people to coordinate their walks home after dark. Meetings were also set up with the mayor's representatives and the head of New Haven's police force. The first round of deliberations led to the posting of a police car after 6:00 p.m. at the corner of the street where most of the muggings occurred. This solution, however, proved only temporarily dissuasive, as the muggings would simply occur when the police car was not there. Another explored solution was to post an undercover agent in the dangerous location in the hope of identifying and catching the criminals. The time being the middle of January, however, this option was not really viable either. After another round of deliberation, somebody suggested installing lights on the bridge, as the darkness of that bridge after sunset invited crime. This simple, commonsense suggestion struck everyone as far superior to the previous solutions, and it quickly garnered a consensus. Unfortunately, however, a technician from city hall then explained that there is a high-voltage system under the bridge, which crosses over a railroad track, making it impossible to use electric lights to light the bridge. As this solution seemed about to be ruled out, someone else asked if this impossibility applied to solar lamps; it did not. Moreover, it was discovered, solar lamps offer the additional advantage of being maintenance free. The city hall accountant, however, pointed out that there was a budget constraint. Solar lamps cost at least $5,000 each and the city simply could not afford them.

Finally, another participant asked whether the city could not ask the federal government for some stimulus money. In the end, the city purchased and installed three solar lamps, for a total of $40,000, paid by federal money.[11] A block party was organized on the bridge by the neighbors in late September to celebrate the installation of the lights. Since then, and as of November 2010, not a single mugging event has been reported in this specific area.[12]

This example illustrates how different approaches to the problem (that of the regular citizens, the police, the engineer, and the accountant) combined to guide the group from the most obvious but suboptimal solution (the police car posted at the corner of the dangerous block) to the less obvious and more compelling solution (solar lamps on the bridge). The example is also meant to illustrate how a group of

[11] For a full report of the story, see http://newhavenindependent.org/index.php/archives/entry/and_wooster_square_said_let_there_be_light_and_there_was/

[12] I base this assessment both on periodic checks of http://seeclickfix.blogspot.com/2010_09_01_archive.html (a website that allows local communities to report problems and work collaboratively on getting them fixed, if only by pressuring the relevant authorities through online petitions) and the e-mail reports I receive weekly from Karri Brady, the neighbor who heads the Wooster Square community.

nonexperts can do better than the experts themselves, in this case, the police. This is because the police kept trying to offer solutions along the dimension they knew best—either catching the muggers or dissuading them by overt presence—when the more compelling solution required thinking outside the experts' box and taking a different approach. Of course, it remains to be seen whether crime truly decreases over the long term—as that will be the only way to validate the chosen solution as meeting the procedure-independent standard of correctness in this particular case. I take it, however, that the chosen policy was, by far, the best of those explored until then.

According to Lu Hong and Scott Page's results on the components of collective intelligence (Hong and Page 2001, 2004; Page 2007), what matters most for the quality of collective problem solving of the type described in the previous examples is "cognitive diversity." Cognitive diversity is the difference in the way people will approach a problem or a question. It denotes more specifically a diversity of perspectives (the way of representing situations and problems), diversity of interpretations (the way of categorizing or partitioning perspectives), diversity of heuristics (the way of generating solutions to problems), and diversity of predictive models (the way of inferring cause and effect) (Page 2007: 7). Cognitive diversity is not diversity of values or goals, which would actually harm the collective effort to solve a problem. Because of the importance of cognitive diversity thus defined, given specific conditions, "a randomly selected collection of problem solvers outperforms a collection of the best individual problem solvers" (Hong and Page 2004: 16388; Page 2007: 163). This is the Diversity Trumps Ability Theorem.

Importantly, the four conditions for this theorem to apply are not utterly demanding. The first one simply requires that the problem be difficult enough, since we do not need a group to solve easy problems. The second condition requires that all problem solvers are relatively smart (or not too dumb). In other words, the members of the group must have local optima that are not too low; otherwise the group would get stuck far from the global optimum. The third condition simply assumes a diversity of local optima such that the intersection of the problem solvers' local optima contains only the global optimum. In other words, the participants think very differently, even though the best solution must be obvious to all of them when they are made to think of it. Finally, the fourth condition requires that the initial population from which the problem solvers are picked must be large and the collection of problem solvers working together must contain more than a handful of problem solvers. This assumption ensures that the randomly picked collection of problem solvers in the larger pool is diverse—and in particular, more cognitively diverse than a collection of the best of the larger pool, which would not necessarily be the case for too small a pool relative to the size of the subset of randomly chosen problem solvers or for too small a subset of problem solvers in absolute terms.[13]

The general point of the Diversity Trumps Ability Theorem is that it is often better to have a group of cognitively diverse people than a group of very smart people who think alike. This is because whereas very smart people sharing local optima will tend to get stuck quickly on their highest local common optimum, a more cognitively diverse group has the possibility of guiding each other beyond that local optimum toward the global optimum. We can imagine that, in the scenario of *Twelve Angry Men*, if the jury had been composed of clones of juror number 8, the smartest person in the lot, they might have been stuck with his initial suspicion but unable to turn it into the firm conviction of "not guilty" reached by the group. Similarly if all three *députés* were thinking exactly alike—say, like the *député* of Calvados who thinks of his lower local optimum first—no matter how long they deliberated, their group would stay stuck on the local optimum of Grenoble (9) and would never be able to reach the higher local optimum of Caen (10). If all thought like the *député* of Calvados or Corrèze, they would still have a given probability of reaching the global optimum, but not the certainty of the deliberating group described above.[14] Finally, if only the police had been involved in the search for a solution to the crime problem of Wooster Square, they would have never thought of solar lamps as the best solution.

It is thus often better to have a group of cognitively diverse people than a group of very smart people who think alike. Notice that to the extent that cognitive diversity is correlated with other forms of diversity, such as gender or ethnic diversity, the argument suggests that affirmative action and the use of quotas might be a good thing not just on fairness grounds but also for epistemic reasons.[15] In the last section of this chapter, however, I will show that it is problematic to assume stable correlations of that kind.

---

[13]Notice that the first part of this fourth condition can be thought of as Madison's requirement in *The Federalist* No. 10 that the pool of candidates to the position of representatives be large enough. For more on the four conditions, see Page 2007: 159–62.

[14]Each member is defined by a set of local optima and a probability of getting stuck at each of his local optima. So if the deliberating group is made up of the exact same people who have a nonzero probability of getting stuck at the nonglobal optimum, the group probability of finding the global optimum might be higher than that of any individual in the group, but it won't be a 100 percent.

[15]For a defense of cognitive diversity as being in fact the "only" reason to support affirmative action, see the conclusions of the French sociologist Sabbagh 2003.

3. Why More-Inclusive Deliberating Groups Are Smarter

Deliberation is not by itself democratic or inclusive. In effect, deliberation can theoretically occur within one person (degenerate case) or among a few oligarchs. The examples I discussed involved at most a few dozen people. What is, then, the advantage of involving large numbers?

I hypothesize, very simply, that the advantage of involving large numbers is that it automatically ensures greater cognitive diversity. In that sense, more is smarter, at least up to the point of deliberative feasibility. I thus propose to generalize the Diversity Trumps Ability Theorem into a Numbers Trump Ability Theorem, according to which, under the right conditions and all things being equal otherwise, what matters most to the collective intelligence of a problem-solving group is not so much individual ability as the number of people in the group. If three *députés* are more cognitively diverse and thus smarter than just one, then five hundred are likely even more cognitively diverse, and thus smarter, than three. Similarly, if twelve jurors are smarter than one, then so would forty-one or 123 jurors. Of course, this assumption that cognitive diversity positively correlates with numbers will not always be verified but it is generally more plausible than the reverse assumption that cognitive diversity increases as the number of people goes down.[16]

What I propose here, in effect, is a hypothesis about the missing link of existing arguments attributing epistemic properties to democratic deliberation. Most such arguments are too broadly tailored, showing a connection between any kind of deliberation and epistemic properties. They do not tell what is special about inclusive versus less inclusive democratic deliberation (e.g., Estlund 2008). Some even tell us that there is nothing special about deliberation with others when contrasted with internal deliberation or "deliberation within" (Goodin and Niemeyer 2003; Goodin 2008, on which more in the next chapter). The missing conceptual link between epistemic properties and inclusive democratic deliberation is, to repeat, the following: larger deliberating groups are simply more likely to be cognitively diverse. To the extent that cognitive diversity is a key ingredient of collective intelligence, and specifically one that matters more than average individual ability, the more inclusive the deliberation process, the smarter the solutions resulting from it should be, overall.

---

[16] A complicating factor is probably the (s)election mechanism. In selecting, say, a hundred representatives, a system of proportional representation may produce more cognitive diversity than majority voting in single-member districts. This invites an epistemic comparison between alternative democratic selection mechanisms, some of which can produce more cognitive diversity with fewer additional members.

4. REPRESENTATION

A crucial problem, however, which is bound to dampen any enthusiasm one might have for numbers, is a question of threshold. Deliberation involving all members of the group is not always feasible, at least not as "simple deliberation" involving individuals talking to each other face-to-face. James Fishkin and Bruce Ackerman (2004) have notoriously argued for the defense of a national holiday that would allow the entire nation to deliberate once a year across smaller subgroups, but even this proposal recognizes the impossibility of having all people deliberate with all others at the same time. On too large a scale, what we are dealing with is the less conceptually tractable "distributed deliberation." Since our focus in this chapter is simple deliberation, we have to assume that there is a point, therefore, beyond which things cannot be equal otherwise, such that including more people will simply worsen the quality of deliberative outcomes even if we assume that it still increases the cognitive diversity of the group.

In a democracy, the institutional device of representation comes as an obvious solution to that problem of threshold. Representation by itself has a historically dubious democratic pedigree and arguably qualifies as an aristocratic institution, because of the principle of "election" of a governing elite with which it is generally associated, leading many interpreters to see representative government as, fundamentally, a "mixed regime" (e.g., Manin 1997). For elitist democrats from Burke to Schumpeter and followers,[17] the function of representation is to select the more competent citizens, who are then in charge of refining citizens' judgments and preferences and making presumably better laws and decisions on their behalf.

Nonetheless, since the eighteenth century, representation has evolved toward more democratic forms. In theory, it can allow the indirect or mediated involvement of the many in a decision made by the few. In other words, representation is an institutional device that has the potential to make democratic decision making possible in a mediated way when numbers are too large. Nadia Urbinati, for example, whose interpretation of representative democracy is resolutely antielitist, argues that the superiority of representative democracy over direct democracy is not to be found in an a priori higher competence of elected representatives but in the reflective wisdom that the feedback loop between representatives and

---

[17]Of course, there are sensible differences between Burke and Schumpeter. On the Burkean view, the representatives are supposed to use their better judgment to make decisions about the general good and the long-term interests of citizens. On the Schumpeterian view, the representatives use their brains to maximize their chances of being reelected by pursuing policy goals likely to satisfy their constituents' preferences.

the represented allows, and the time delay that this introduces between preferences and decisions (Urbinati 2006).

For all their differences, a common point between the elitist and democratic approach to representation presented here is to see representative democracy not simply as second best to direct democracy, rendered necessary by the feasibility constraint, but as a way to improve on the decisions that regular citizens would make by delegating the task to professional politicians. The epistemic improvement introduced by the device of representation can thus be, on these existing views, twofold: first, representation brings to power the more capable citizens; second, it introduces the possibility of a paced and reflexive thinking process in the back-and-forth between the people's input and the proposals of the representative assembly (Urbinati 2006).

The argument from collective intelligence suggests, however, another democratic way to look at representation and the function fulfilled by representatives. In this view, the primary function of representation is not to gather the best and brightest, nor is it simply to introduce a reflective dimension into the collective decision-making process (although it might be that too). Here, the function of representation is to reproduce the cognitive diversity present in the larger group on a scale at which simple deliberation remains feasible. From the epistemic point of view, the expectation remains that the representative assembly should, among other things, come up with the best possible answers to all sorts of collective problems. However, the assumption is that this epistemic performance is primarily conditional on the systemic properties of the assembly, including its cognitive diversity, rather than on the individual abilities of its members.

In order to see how an assembly of elected representatives may be smarter than an oligarchy of the wise, let us compare the solution of an assembly of, say, five hundred congressmen with the oligarchic solution of five hundred self-selected individuals.[18] Bracket for now the question of motivation and assume that all these individuals are motivated to act in the interest of the ruled. Imagine further, for the sake of argument, that the five hundred oligarchs in our example are extremely smart and knowledgeable. Let us finally add that these five hundred oligarchs are at least as cognitively diverse as a group as the five hundred congressmen. How could we not prefer the oligarchy of five hundred such individuals over an assembly made up of individuals chosen by regular citizens?

The reason is that the ideal characteristics of the oligarchic assembly posited here are not just highly implausible; they can hardly be expected

[18] Those numbers approximate the reality of the Republic of Venice in the fifteenth century, which was governed by a few hundred aristocrats.

to be maintained over time. This is because of at least two factors. One is that absent periodic renewal of their members (through both limited terms and a limited number of terms), the group of oligarchs is likely to be stuck with a given level and type of cognitive diversity. This might not be a problem if this initial level is high and political circumstances do not change dramatically—but political circumstances often do change, and often change dramatically. But second, absent democratic account-ability and some form of influence of the larger public opinion on their decision process, the group of oligarchs has no incentive to maintain this initial degree of cognitive diversity, let alone reflect the changing ways of looking at the world within the larger population. In effect, no matter how epistemically superior the group of oligarchs is at the begin-ning, it is unlikely to remain so over time. The group of oligarchs may be characterized by high average political IQ, but ultimately it is likely that they won't be able to maintain enough cognitive diversity within their group. Their epistemic performance can thus be expected to diminish over time.

In the case of the assembly of five hundred congressmen, however, its cognitive diversity can be assumed to be preserved over the long run thanks to periodic elections that renew the pool of members. Further, an elected and accountable assembly, which is at least minimally shaped by a larger public opinion, is more likely to stay cognitively diverse than a body of oligarchs that can only count on the discipline of its members to avoid the trap of groupthink, self-serving biases, and isolation from pop-ular input. So while there might be times when a large enough oligarchy might temporarily be epistemically equal, and perhaps even superior, to a democratic assembly of representatives, over the long run, this is unlikely to remain true.[19]

On that reading of representation, the epistemic argument for delib-eration among the many presented earlier translates from direct to repre-sentative democracy, provided representation reproduces at least some of the cognitive diversity of the larger group in the smaller one. The claim remains the same: deliberation involving the many, in a direct form where feasible or an indirect form through representation, is epistemically supe-rior to deliberation among the few: to the extent that cognitive diver-sity is correlated with numbers and provided that citizens are at least

[19] A contemporary example of an oligarchy whose epistemic success arguably compares with those of democratic regimes, at least as far as economic policies and public education are concerned, would be the Communist regime in China. Because the label "communist" in fact now covers ideological positions ranging from the far right to the far left—the only common ideology being nationalism—one can argue that the policies pursued in China compare with those that would be produced by the (democratic) rule of the median voter. I owe this provocative suggestion to Pasquale Pasquino.

moderately smart on average, the more numerous the deliberating group, the smarter.

I have so far assumed that the default mode of selection for representatives is election by the citizens. Isn't there a way in which we could further enhance the cognitive diversity of the representative assembly, even if that means giving up on the principle of election? I will now argue that random selection would probably do better.

5. ELECTION VERSUS RANDOM SELECTION

Elections as we know them are perhaps not the best way of ensuring the reproduction of the cognitive diversity of the larger group on the smaller scale at which deliberation remains feasible (although it is possible that, under some circumstances, elections do a good enough job). In practice, elections tend to bring to power socially and economically homogenous people, suggesting that the assembly is not likely to be as cognitively diverse as it should.[20] Even in theory, though, it is not clear that the principle of election can be fully reconciled with the goal of cognitive diversity, as there might be a selection bias in the pool of people likely to run for office in the first place. Those people are likely to share some personality traits (a type A personality, say) or other characteristics that may reduce the overall cognitive diversity of the assembly. As a consequence, even if the individual ability of elected representative assemblies is high, their cognitive diversity is unlikely to be as high as we could wish.

Assuming that on average, the citizens from among which the representatives are selected meet a minimal threshold of individual competence, what would be an alternative selection method ensuring as much cognitive diversity as possible in the representative assembly? It depends, of course, on what we mean by "as much cognitive diversity as possible." If the goal is to at least replicate on the smaller scale of the representative assembly the cognitive diversity one would get by including everyone—that is, the cognitive diversity existing at the level of the larger group—an obvious solution is the use of random lotteries. Random lotteries would indeed produce what is known as a "descriptive representation" of the people (Pitkin 1967), or in John Adams's famous formula, "an exact portrait, in miniature, of the people at large" (Adams [1851] 1856: 194–95), ensuring a statistical similarity of thoughts and preferences of the rulers and the ruled.

---

[20] For a compelling critique of and solution to the problems of representative democracy in America, see O'Leary 2006.

Random lotteries have been recently explored as an alternative to elections on many grounds: equality, fairness, representativeness, anticorruption potential, protection against conflict and domination, avoidance of preference aggregation problems, and cost efficiency, among others (e.g., Elster 1989b: 78–103; Mulgan 1984: 539–60 and Mulgan 2011; Goodwin 1992; Carson and Martin 1999; Duxbury 1999; Stone 2007, 2009, 2011; Sintomer 2007; Engelstad 2011; Mueller, Tollison, and Willet 2011; Knag 2011). The descriptive representation that lotteries would achieve, however, is normatively desirable for specifically epistemic reasons as well. Descriptive representation achieved through random lotteries would not elevate the level of individual ability in the deliberative assembly, as by definition the expected individual ability of the selected individuals would necessarily be average, but it would preserve the cognitive diversity of the larger group.

Empirical evidence partially backs up the prediction that under some nonexacting conditions, groups of average citizens perform decently well when placed in the right deliberative conditions. Deliberative polls or citizens' assemblies, for example, which gather between one hundred and five hundred randomly or quasi-randomly selected participants would seem to offer such deliberative conditions. Deliberative polls, despite taking place over two days or fewer, provide participants with briefing material that they can discuss in smaller groups of fifteen or so, as well as access to expert panels that they can question at length during plenary sessions that also have a deliberative dimension. The model of citizens' assemblies, which have a longer life span of several weeks to several months, allows for even more in-depth pre-deliberation reading and processing of information, as well as the pursuit of the deliberative process over many meetings.

Both experiments have been documented to produce epistemically promising outcomes. Deliberative polls (see Fishkin 2009) thus produce more informed postdeliberation preferences on topics ranging from the selection of a candidate for mayor (in the 2006 deliberative poll organized in Greece), energy policies (American deliberative polls), or the reform of the European Union pension system (the 2007 deliberative poll called "Tomorrow's Europe"). Similarly, the 2004 British Columbia citizens' assembly produced, over a period of several months, a sophisticated and innovative proposal on the complex topic of electoral reform meant to address the problem of democratic deficits in Canada (Warren and Pearse 2008). The size of such experiments, involving a little more than one hundred people in the case of citizens' assemblies and up to five hundred in the case of deliberative polls, is sufficiently large to make the random selection truly representative and allows us to extrapolate from the performance of regular citizens in these deliberative contexts to what the epistemic performance of an actual parliament based on random selection would be like.[21]

The results of these experiments should also assuage the fear that the benefits of cognitive diversity would necessarily be offset by increased communication costs and disruptive value diversity in a randomly selected representative assembly. Deliberative polls, in particular, have been conducted with great success across the globe, sometimes despite challenging communicative contexts induced by language barriers, cultural differences, or even profound value rifts, as in the case of the 2007 deliberative poll in Northern Ireland involving Protestants and Catholics (see Fishkin 2009: 159–69; Farrar et al. 2010).

Now, is reproducing the cognitive diversity of the larger group in the context of the smaller group of representatives the best we can do? If the goal is to maximize the cognitive diversity of representative assemblies, an even better method than random lotteries, which simply reproduces in the smaller group the diversity existing in the larger group, would seem to be to oversample the cognitive minorities existing in the larger group. We might thus be better off with a smaller pool of representatives carefully selected for their cognitive differences than with a larger pool of randomly selected representatives. This argument thus puts in question the claim that, at the level of representative assemblies at least, the "more is smarter" claim, however mediated by representation, still applies.

I will argue that the option of oversampling cognitive minorities, however appealing on the face of it, is both impractical and undesirable. If the group were very small, say the size of a jury, it would make sense to try to engineer cognitive diversity consciously since random selection in that case would not guarantee enough of it (because the law of large numbers fails to apply in small samples). On the larger scale of an assembly of five hundred representatives or more, chosen from an even larger pool of candidates, the attempt to pick and choose the relevant kinds of people needed to maximize the problem-solving properties of the assembly is unnecessary and potentially counterproductive.

This is due, essentially, to the unpredictable nature of political questions. Politics is arguably the domain of questions where we collectively deal with the unknown. As a result, it is impossible to identify in advance

---

[21] The results observed in the smaller context of consensus conferences, citizens' juries, and the like, which gather only a few dozen citizens at a time, also support the epistemic properties of deliberative groups, but it is harder to extrapolate from the performance of these smaller groups to the performance of large assemblies. The fact that these groups are self-selected rather than randomly selected also makes them a less scientific source of evidence (e.g., Mansbridge 2010a for a critique).

all the questions that any given representative will have to deal with over the several years of his or her tenure (e.g., a foreign war, a financial crisis, global warming, or terrorist attacks. If we knew in advance what the problems were going to be over the next few years—say problems related strictly to an economic crisis—we would want to ensure that the assembly has enough cognitive diversity along a certain dimension of relevance for economic decisions. We would presumably be better off with an assembly of economically savvy and ideologically diverse representatives, some of whom are Keynesians, some monetarists, and some Austrian-school followers when it comes to macroeconomic principles. We would then have, presumably, enough cognitive diversity of the right kind to increase the chances that deliberations among these representatives produce good outcomes. In fact, if we could know, or at least guess with enough accuracy, what problems the assembly would have to deal with, we should generally oversample deviant perspectives. For example, if we knew that an ecological disaster was likely to take place within the next five years, we could make sure that a legislature contains more environment-friendly individuals than are present proportionally in the overall population.

Whether we can ensure such oversampling in a democratic fashion is not obvious. But that question of legitimacy is preempted, in any case, by a feasibility question. In most cases, we cannot predict the relevant dimension of cognitive diversity in advance, because we simply cannot predict the future. Nor do we know what the relevant cognitive diversity really translates into, as categories such as Keynesian, monetarist, Austrian, or various styles of "environmentally friendly" are very often too crude to capture the differences between people's ways of thinking about economic or environmental issues. Even if a refined ex post facto sociological analysis could reliably correlate certain features with certain views (e.g., left-libertarian females tend to think X on issues of Z), that still might not tell us anything of relevance about what their views would be in the future, in a different context, and on ever-changing issues. Political problems, I surmise, are unpredictable issues for which we cannot tell in advance who is going to have the relevant perspective. The rational attitude to have with respect to such questions is one of agnosticism as to who has the best answer to them, until that answer is tried in the public forum. The only thing we can tell about political problems is that solutions can come from anywhere and are unlikely to come always from the same people, that is, people who can be identified as belonging to specific categories (e.g., whites, males, Republicans).

Combining the uncertainty about the problems an assembly will have to solve in the few years of its tenure and the almost infinite diversity of human cognitive properties makes it technically impossible to implement oversampling.[22] We simply can't tell in advance from which part of the *demos* the right kind of ideas are going to come.[23] It therefore does not make much sense to try to engineer cognitive diversity *ex ante*.[24]

Finally, even assuming that one could identify in advance what kind of cognitive traits matters most to increasing the quality of deliberation on a specific question likely to be on the agenda—economic and environmental issues are, after all, very likely to arise—nothing guarantees that these traits will be equally relevant on other questions. Worse, any attempt at oversampling cognitive minorities on the basis of classical statistical categories is bound to homogenize the representative assembly along one or more dimensions, which actually risks harming the epistemic potential of the deliberating group. Assuming, for example, that you oversample economically savvy individuals or environmentalists to make sure the deliberation on some economic or environment-related issue is the best it can be, you have no guarantee that oversampling along those lines has preserved cognitive diversity when it comes to addressing entirely different issues. In fact, by oversampling any category of people, you may have unknowingly homogenized the group of representatives along lines that are very problematic for other issues. Economically savvy individuals, however cognitively diverse with respect to economic issues, may all tend to be too lenient toward Wall Street and financial institutions and corporations in general. Environmentally friendly individuals of different styles may tend to be too fiscally irresponsible. (I hope I'm making up both examples.) On some other issues—say, the problems related to poor single mothers in black communities—a group in which both economists and environmentally friendly individuals are disproportionately represented may well lack enough of the relevant diversity because oversampling economics PhDs and environmentally friendly citizens may mean oversampling white privileged individuals. Random sampling is the

---

[22] Again, things would be different if we were talking about small deliberating groups of, say twelve people or fewer, for which it might make sense to draw cognitive style distinctions.

[23] Not to mention the fact that, as we saw in the stylized examples presented earlier in the chapter, it is not always the case that the "right" idea simply "comes from" one individual or part of society. The emergence of the "right" idea or solution is a product of a group dynamic, in which someone's suggestion unpredictably triggers someone else's input—as when one *député* suggests Marseilles because it is a big city, which provokes another to argue for Paris, and so on. It is also impossible to preselect individuals for this kind of dynamic.

[24] See chapter 3 for a defense of the idea that the belief in the unpredictability of politics might well account for the Greeks' principle of *isegoria*. In politics, it is better to let everyone speak in the assembly, because, unlike what happens in more technical domains like architecture or shipbuilding, we simply don't know in advance who will come up with the answers or who will bring the relevant perspectives and arguments.

simplest, most parsimonious way to avoid this kind of bias and get as much cognitive diversity as possible in the absence of knowledge about the kinds of perspectives that will ultimately be needed.

Since an assembly cannot be reconfigured at will for every possible new issue, and since every issue may require a completely different type of cognitive diversity, in the end it seems more rational to consider each person in the group a unique source of potential cognitive diversity and to try to preserve in the legislature the many unique perspectives of the larger group.[25] Another solution would be, perhaps, to convene a different assembly for every possible issue, thus fragmenting the decision-making process and distributing it over many specialized assemblies. I will not entertain this interesting solution here, as I assume throughout that we want to preserve some of the centralizing features of existing representative assemblies. We can note, however, that even Fishkin's deliberative polls, which specify in advance the issues to be discussed, are designed around the principle of random selection as preferable to that of minority oversampling. This holds even when minorities, defined as communities of interest, are historically well defined, as in the case of Catholics versus Protestants or Bulgarians versus Roma (Fishkin 2009: 163).[26]

An objector might target the use of lotteries as preferable to elections on other epistemic grounds. Even assuming that cognitive diversity matters indeed more than individual ability for the problem-solving abilities of the representative assembly, lotteries might lower the average ability far below the threshold necessary for group competence, so that we would still be better off with a homogeneously thinking group of representatives rather than a representative group of average citizens. In other words, the average citizen might be simply too dumb. In that case, shouldn't we privilege instead a form of "selective descriptive representation" that would consist in choosing among various predefined categories of people

[25] It is true that, in practice, if people can't be forced to participate and pure randomness must be abandoned, then some consideration for quotas based on gender, ethnicity, or other rough-and-ready categories might be better than nothing. This is the solution applied in the quasi-random recruitment of participants to citizens' assemblies, for example (Warren and Pearse 2008). But on the general abstract principle, the fact remains that systematic oversampling of cognitive minorities is both unfeasible and generally normatively undesirable for epistemic reasons.

[26] One notable exception is the 2001 deliberative poll organized in Australia on the fate of Aborigines (Fishkin 2009: 162), where the massive disproportion between the Aborigines, who represent less than 3 percent of the population, were oversampled in proportion to the rest of the population to make sure they would form a critical mass in the final deliberative sample. Still, I would argue that this scenario (where both the problem to be discussed is known in advance and a sufficiently small, stable, and relevant minority can be equated with a cognitive minority with respect to the topic at hand and can be identified *ex ante*) does not characterize the situation facing a normal representative assembly.

those that are deemed smarter than the rest?[27] Combining elections with quotas or the right kind of district design, for example, could thus get us the best of both worlds: elected individuals with high individual abilities and some of the cognitive diversity present in the group at large.

Let us grant, for the sake of argument, the optimistic assumption that elections do, in fact, select the best and brightest problem solvers, or that such political "experts" exist at all (see Tetlock 2005 for skepticism as to the superiority of experts' political judgments over those of laypeople[28]). First, it still remains unproven that the average citizen's individual abilities are below the threshold that must be met for competent problem solving in a sufficiently cognitively diverse group. If anything, the available empirical evidence speaks in favor of the opposite conclusion. Again, the results observed in deliberative polls or citizens' assemblies seem to falsify the theoretical worry about the inaptitude of the average citizen for figuring out solutions to complex questions with others. If so, there might be (at worst) a strict tie between a cognitively diverse group of individuals with average individual abilities and a group of homogenously thinking people with high individual abilities, even when some amount of cognitive diversity has been injected in the latter group through quotas or gerrymandering.

Second, as already said, quotas and gerrymandering are extremely clumsy and imperfect ways to inject cognitive diversity in the representative assembly. While it may be the case that the sociological features on which quotas and gerrymandering are based are going to be correlated with the right kind of cognitive difference for some problems (e.g, that brought by women or blacks), the relevant categories for other problems would need to be something else entirely (e.g., animal lovers or librarians). For most political situations, we cannot know in advance which property of the electorate will be relevant—that is, from which category of people the right kind of thinking will come.

Finally, it is far from certain that the result of elections with such correcting measures for selective descriptive representation would be much different in terms of cognitive diversity than elections without such measures. After all, as pointed out earlier, if it is the case that elections draw a certain type of person (e.g., Type A personalities), diversifying the group

[27] Selective descriptive representation is currently achieved in several European countries through a system of quotas ensuring the presence of various minorities on party lists, and in the United States by gerrymandering new districts to ensure the election of representatives of those same minorities. Some authors advocate selective descriptive representation in certain contexts and for historically disadvantaged groups. Jane Mansbridge's argument for it, for example, is that selective descriptive representation can enhance the substantive representation of these disadvantaged groups' interests as well as improve the self-image of those communities or increase the polity's de facto legitimacy (Mansbridge 1999: 1).

[28] When it comes to assessing a problem and making political predictions, Tetlock argues, political "experts" hardly do better than laypeople and, on the purely predictive side, are in general outperformed by simple statistical regressions.

along some dimensions (e.g., ethnicity, gender, religion) while retaining that common trait might not necessarily do much good (although it is probably better than nothing).

As a result, it seems that the best solution is not to choose but to leave it up to chance and the law of large numbers. Trying to predict whether a black single mother or a Caucasian farmer would contribute more to the quality of the deliberative outcome on a series of topics yet to be determined is silly at best and essentializing at worst. From an epistemic point of view, therefore, selective descriptive representation (particularly if based solely on a history of past discrimination) is not likely to create enough cognitive diversity. There may be valid reasons to embrace gerrymandering and quotas, but it is doubtful that improving the epistemic properties of the deliberating assembly is one of them.

There is, arguably, one powerful reason to worry about random selection, and perhaps to prefer instead some selective descriptive representation scheme. The concern about random selection is not that randomly selected citizens would be of excessively low average intelligence but rather that this selection mode would one day lead to the appointment of extremely incompetent and/or morally corrupt individuals (e.g., Nazis or white supremacists) who would cause important kinds of problems (epistemic and otherwise). Over time, indeed, under a continuous system of unrestricted random sampling, the probability of such an unlucky draw goes to 1. This is, however, a very theoretical worry. Consider even the pessimistic scenario of a population where 25 percent of the population consists of these really incompetent people—let us identify them as "white supremacists"—and we aim to randomly appoint an assembly of, say, fifty representatives, to be renewed every four years.[29] The first time we use the random sampling mechanism, the probability of drawing an assembly in which there is at least a simple majority of white supremacists (that is, twenty-six of them or more) is ridiculously low: 0.0038 percent. Over time, however, as we keep using the procedure, this probability will, as the objection points out, rise to 100 percent. This will happen, however, over an *infinite* amount of time. How many years would it take for this probability to rise not to 100 percent but, say, 50 percent? The answer is 72,924 years. For the risk to go up only to 10 percent, we would still have to wait 11,088 years. For the risk to rise to 1 percent, it would take 1,060 years. No democracy has lived that long, and at least some representative

---

[29] The calculus grows more unwieldy as the size of the assembly increases, hence the choice of that relatively low number. The point is, in any case, that it would take even more time for the probability of drawing a "bad" assembly consisting of several hundred individuals to reach any dangerous threshold, so the argument that follows applies a fortiori to the case of most existing representative democracies, whose representative assemblies are generally ten times as numerous as the example considered here.

democracies based on the election principle have managed to produce much worse assemblies in much shorter periods of time.

It is true that we could be terribly unlucky and, against the odds, draw the dangerous assembly on the first trial or soon after. In a well-designed democracy, however, there should be institutional safeguards that limit the damage potentially caused by a particularly bad, if unlikely, draw. Constitutional checks and the existence of a second, nonrandomly selected chamber, for example, may come to mind. While the objection thus raises a genuine, though highly theoretical, problem, when this problem is weighted against the potential benefits of random sampling, it does not seem to be enough to justify throwing out the baby with the bathwater (see also, for a refutation of the same objection along similar lines, Mueller et al. in Stone 2011: 54).

Let me, finally, consider a more specific objection to my defense of random selection of representatives that can be built on Andrew Rehfeld's bold proposal to keep the election principle but randomize constituencies (Rehfeld 2005). Rehfeld provocatively suggests randomly assigning for life every new voter, upon their registration at age eighteen, to one of 435 virtual constituencies. The goal of this reform would be to create stable, heterogeneous, and involuntary constituencies that would form small mirror images of the nation, rather than interests or identity-based districts as is currently the norm with territorial districts. According to Rehfeld, this reform would not only bring us closer to the real intentions of the Founding Fathers when they designed large territorial districts but also closer to the normative ideal of legitimate representation. What is particularly interesting for our purposes here is that, according to Rehfeld, mirror-image districts would foster truly common-good-oriented voting, as opposed to voting oriented toward the defense of local interests, simply in virtue of the fact that the good of representatives' constituents would be identical to the good of the whole.

Notice though that Rehfeld's randomized constituencies are problematic from an epistemic point of view. To the extent that constituencies are randomized, they are all like miniature versions of the people. As such, the way they will vote is most likely going to be quite similar. This similarity, according to Rehfeld, guarantees that random constituents will vote for the common good, which is statistically equivalent to the good of any random constituency, rather than now-nonexistent local interests. From this point of view, there is no doubt that random constituencies are a marked epistemic improvement on the practice of gerrymandering in terms of selecting the "right" representative.

However, at the collective level, this epistemic advantage is largely annulled by the fact that taken as a group, those "right" candidates do not add up to the "right" assembly. Since statistically similar constituencies would presumably vote for the same type of person most of the time, the US Congress would end up being entirely made up of white male Republicans. Randomized constituencies may improve the civic-mindedness and "rightness" of the votes, but they would also likely lead to a cognitively homogenous assembly of 435 representatives who think roughly the same way.

Rehfeld (2005, epilogue) considers the objection that the first Congress after the transition to random constituencies that he imagines would be heavily homogeneous. However, he trusts that as far as the ideological imbalance is concerned, at least, things would smooth over after some time. After an all-Republican Congress, the next round of elections would bring in an all-Democratic Congress. After a few more oscillations, the Congress would stabilize in the middle of the ideological spectrum, as the importance of partisanship would diminish over time. Regarding the problem of minority voices, Rehfeld's solution consists in reintroducing "quotas" at the level of random constituencies. By changing the qualifications for holding a congressional seat and making it so that, for example, only African Americans or women could run for election in a certain number of constituencies, he argues that one can reintroduce some degree of diversity at the level of the representative assembly in a controlled and transparent manner. While the proposal is a marked improvement on gerrymandering practices that homogenize districts along racial lines (not gender ones though) to a comparable effect, the problem from an epistemic point of view is that such quotas suppose that one can determine in advance which kind of diversity (race, gender, and so on) is good for deliberation. As already argued, nothing guarantees that this predefined diversity exactly maps onto the epistemically optimal cognitive diversity.

This chapter has argued that deliberation inclusive of all the members of the demos (however the latter is defined) can be expected to have greater epistemic properties than less inclusive deliberation because of the greater cognitive diversity a more numerous group is likely to tap. This claim is meant to hold whether deliberation involves all the members of the group directly, where feasible, or indirectly, when the group is too large and some form of delegation to representatives must take place. In the case of indirect or delegated deliberation, a counterintuitive implication of the importance of cognitive diversity for the epistemic performance of a deliberating group is that random selection is preferable to elections as a selection mechanism for representatives. Random selection indeed preserves the cognitive diversity of the larger group, whereas elections will tend to reduce it.

The arguments presented here have proceeded with a somewhat idealized notion of deliberation. It is now time to turn to the various problems that are known to affect real life discursive exchanges and assess the extent to which they threaten the epistemic claim in favor of inclusive deliberation.

CHAPTER FIVE

Epistemic Failures of Deliberation

IN THIS CHAPTER, I turn to a series of objections that can be generally addressed to the ideal of deliberative democracy and count more specifically as objections to the claim that deliberation, and particularly inclusive, democratic deliberation, has epistemic properties.

The ideal of democratic deliberation at the heart of the previous chapter has been criticized from many fronts since its first formulation in the late 1980s. As a normative ideal, it is often attacked for being too demanding and too utopian to be worth pursuing. At one extreme, critics argue that democratic deliberation, in practice, does not do much to change people's minds. At the other extreme, others argue that it changes them for the worse. Robert Goodin and Simon J. Niemeyer (Goodin and Niemeyer 2003; Goodin 2008: chap. 3), for example, worry that as a dialogical exchange between citizens, democratic deliberation may do less to change people's preferences than the kind of monological reflection that they advocate instead as "deliberation within" (Goodin 2000, 2003, 2008). By contrast, Cass Sunstein has repeatedly warned of the dangers of deliberation given what he calls the "law of group polarization," or the tendency of some groups to move toward a more extreme version of the group's pre-deliberative preferences (Sunstein 2002). Given the extent to which he believes that this law afflicts groups, Sunstein suggests (as does Surowiecki 2004) that we should often privilege nondeliberative judgment aggregation over deliberation (see also Sunstein 2007).

Empirical results in political science and social psychology seem to buttress this skeptical view of the normative ideal of democratic deliberation at the heart of deliberative democracy. Sometimes group deliberation homogenizes attitudes, sometimes it polarizes them; sometimes group decisions are better than individual decisions, sometimes not (e.g., Kerr, MacCoun, and Kramer 1996). When it comes to evaluating what democratic deliberation does and whether it does anything good, "the general conclusion of surveys of the empirical research so far is that taken together the findings are mixed or inconclusive" (Thompson 2008). Even the results observed in James Fishkin's deliberative polls, conducted with success across the globe, including in societies divided along religious or linguistic lines (e.g., Fishkin 2009), do not entirely settle the question. Deliberative polls measure postdeliberative preferences and opinions against pre-deliberative ones, without framing the measurement in epistemic terms or providing more than negative reasons to believe that the postdeliberative views actually converge toward the truth. At best, one can verify that the beliefs associated with the new preferences and opinions are more informed and that the usual suspected reasons for poor epistemic outcomes (group polarization, lack of diversity, lack of single peakedness, etc.) do not afflict deliberations.[1]

Few democratic theorists, myself included, would be willing to claim that the normative ideal of democratic deliberation can remain immune to empirical challenges. Most of them are or ought to be bothered by the remaining uncertainty regarding the epistemic properties of democratic deliberation, particularly on salient issues.

This chapter follows two strategies to assuage some of the doubts raised by this empirical literature. The first consists in acknowledging the problems, putting their importance with respect to the general epistemic claim in favor of democracy in the proper perspective and, where necessary, proposing or rehearsing possible pragmatic solutions to them. This is what I attempt in the first section.

I pursue another strategy in section 2, where I rely on a joint work with psychologist Hugo Mercier to offer a theoretical response to the empirical challenge to the epistemic properties of deliberation. This theoretical response consists in using a new psychological theory of reasoning—the "argumentative theory of reasoning"—to make sense of the successes and failures of empirically observed deliberations in a way that rescues the theoretical claim that deliberation has epistemic properties. This argumentative theory of reasoning, I will argue, also allows us to refute or at least take the edge off of some of the most damning critiques of deliberation, namely that it does not do much to change people's minds or, conversely, that it only changes them for the worse. The theory also makes sense of individuals' notorious "confirmation bias"—a bias that persists even in the nonpartisan context of deliberators sincerely aiming for the truth.

---

[1] See also Farrar et al. (2010) for a recent effort at identifying more precisely what does the transformative work in deliberative polls. The results presented in that article go some way toward demonstrating that deliberation has epistemic properties, but they are far from conclusive. Indeed, what the results show is that something happens during the formal, face-to-face deliberation, in contrast with the informal deliberation phase that precedes it. The study, however, does not really open the black box of deliberation per se. Furthermore, it remains to be shown that more informed opinions contribute to better judgments overall, as measured against a procedure-independent standard of correctness that is not purely factual. While it is indeed likely that more informed opinions correlate with better political judgments, this is not necessary. In order to verify this assumption, the experiments would have to be framed in explicitly epistemic terms rather than in terms of measuring a variation in pre-deliberative and post-deliberative opinions.

1. General Problems and Classical Solutions

In practice, the main cause that can negatively affect the quality of deliberative outputs is what is called "social influence" (Sunstein 2002). Social influence includes two aspects, "informational" and "reputational." On the informational side, it is well known that the statements and acts of some group members convey relevant information, and that this information often leads other people not to disclose what they know because they think it is irrelevant. On the reputational side, social pressures imposed by some group members often lead other group members to silence themselves because of fear of disapproval and associated harms (Sunstein 2006: 65–69). In the film *Twelve Angry Men*, to return to the example developed in the previous chapter, the bully juror (juror number 5) exercises this kind of social pressure on the others, some of whom would not have spoken their mind had juror number 8 not countered this pressure through personal encouragements.

These two phenomena—informational and reputational influence—lead to the propagation of errors, hidden profiles (a problem arising when the group slights information held by only a few people), cascade effects, and group polarization. The attempt of Sunstein and his colleagues to implement a type of "Deliberation Day" in Colorado illustrates the latter phenomenon of group polarization. Because participants in this event were separated into towns (one conservative, the other liberal), they were sorted into groups of primarily like-minded people. At the day's end, a narrowing consensus did occur in each group, but the liberals had become more liberal than before deliberation and the conservative groups had become more conservative.[2]

In general, experiments show that as a result of these collective biases, groups tend to examine fewer alternatives, gather information selectively, create pressure to conform within the group, or withhold criticism and collective rationalization. An example is the 2004 report of the Senate Select Committee on Intelligence, which explicitly accused the Central Intelligence Agency (CIA) of groupthink, whereby the agency's predisposition to find a serious threat from Iraq led it to fail to explore alternative possibilities or to obtain and use the information that it actually held (Sunstein 2006: chap. 2).

These well-known problems resulting from group dynamics are compounded by individual cognitive limitations and biases. Those individual biases can be introduced not only by sources of "hot irrationality" (Elster

---

[2] See Cass Sunstein, "Deliberation Day and Political Extremism," University of Chicago Law School Faculty Blog, http://uchicagolaw.typepad.com/faculty/2006/02/deliberation_da.html (accessed February 2, 2006).

2007: chap. 2), such as passions and strong emotions, but also sources of "cold irrationality," such as the heuristics and biases that constrain our judgments. Since Kahneman and Tversky's seminal work (Kahneman and Tversky 1973; Kahneman, Slovic, and Tversky 1982), psychologists have identified all sorts of individual heuristics that individuals resort to when making decisions, some of which tend to systematically lead them astray. On factual matters, the answers to which can be known in advance and independently, this is particularly easy to demonstrate, but one can suspect that the same applies for other matters as well. Examples of such problematic heuristics are the "vividness" and "availability" heuristics, by which people tend to give more weight to the most striking pieces of information or simply to those pieces of information that they have, instead of looking for information that is lacking but would be more relevant.

The film *Twelve Angry Men*, again, nicely illustrates both hot and cold irrationality as well as the more basic problem of perceptual limitations of human beings, who famously make for poor eyewitness. An example of perceptual bias in the film is illustrated by the fact that few jurors noticed the marks on the nose of one of the key witnesses, indicating that she was wearing glasses shortly before being called to the bar (and that her testimony is not as trustworthy as it first sounded, when no one knew she was nearsighted). An example of emotional bias, or "hot irrationality," is given by the angry reaction of juror number 3, a distraught father who projects his failed relationship with his son on the young suspect. Examples of cognitive biases induced by "cold irrationality" are given by the fact, in the Wooster Square example, that the police officers participating in the deliberation could not think of solutions to the crime problem outside of the paradigm of catching the muggers or dissuading them through police-related means. The fact that, in both examples, collective reason eventually triumphs in spite of these pitfalls is not a guarantee that democratic deliberation is always immune to them.

To top it all off, it seems to be the case that individuals are hardwired in such a way that changing one's mind is costly. Contrary to what the theory of deliberation aiming for rational consensus presupposes, the force of the better argument does not always triumph without a fight, because people care about their false beliefs and will long resist amending them. Relying on a vast psychological literature, Mackie (2006) shows that when encountering an argument that challenges their beliefs, people's first reaction is one of denial and strong rejection, no matter how well supported the offending view is. To the extent that every challenged belief is enmeshed in a net of other beliefs that it is cognitively costly to rearrange so as to make room for a new or modified belief, deliberation may well at first stiffen the believer's attitude and polarize his views. The nervous breakdown that juror number 3 has to go through in the film before finally yielding to the irresistible truth illustrates the psychological cost of giving up one's beliefs.

These findings need not prove entirely dispiriting, although I will need to appeal to a much broader theory of individual reasoning to prop up a more general defense of deliberation—a task I turn to in the next section. For now, let me simply remark, as others have, that collective and individual cognitive failures can at least partially be fixed or compensated for by designing the deliberative setting the right way, such that it accommodates the known cognitive biases of human beings. It is clear that the framing of questions and the priming of participants matter in the quality of both deliberative processes and outputs. Fishkin's deliberative polls—those groups of citizens formed by random sampling of the larger population—did not suffer from group radicalization, apparently because people were given balanced, accessible briefing material and were truly exposed to conflicting viewpoints, not just a variety of viewpoints. Pros and cons that were laid out clearly helped people identify the better argument, just as Habermas would have hoped. The ideal of epistemically reliable deliberation is not so unattainable when the right conditions are in place—for example, when the structure of the deliberative setting allows for the neutralization or limitation of the informational and reputational aspects of social influence.

Another way to enhance the epistemic properties of deliberation is to develop formalized methods that challenge assumptions and groupthink—methods such as "red teaming,"[3] "devil's advocacy,"[4] and other types of alternative or competitive analysis. These techniques consist, in essence, of reintroducing some cognitive diversity where it is lacking or stifled.

As to the question of whether or not deliberation can change minds, I will address this problem at length in the next section, but it is worth noting that the reluctance to revise one's beliefs may simply entail that a change of mind will take a long time, at least on some issues. It does not

---

[3]In the US army, "red teaming" is a process executed by members of a given team that provides commanders an independent capability to continuously challenge plans, operations, concepts, organizations, and capabilities. The members of the red teams are educated to look at problems from the perspectives of the adversary and the United States' multinational partners, with the goal of identifying alternative strategies. Some of the team's responsibilities are to challenge planning assumptions and conduct independent analysis in examining courses of action so as to be able to identify vulnerabilities.

[4]Devil's advocacy is the institutionalization of dissent in the context of deliberative decision making, when a person or a team is assigned the task of making the strongest possible case for the counterarguments to the majority's views or even defend the position that no one would defend otherwise.

necessarily mean that a change of mind is ultimately impossible or that a reluctance toward belief revision is ubiquitous on low-stakes issues. In the long term, even on high-stakes issues, beliefs may weaken up to the point where they suddenly die and fall apart, letting an entirely different viewpoint, sometimes the opposite of what was previously so fiercely held, take their place. From the outside, and in the short term, postdeliberative beliefs may seem unchanged, if not reinforced. It might be, however, and in fact it often is the case, that sometime after the exchange of arguments has taken place, beliefs suddenly change and the individual gives in to the new point of view. Gerry Mackie thus compares this process to the way the Berlin Wall fell overnight, for no apparent reason but the slow and irresistible sapping of Communist ideology, creating a collective change of mind that was as sudden as it was unpredictable (Mackie 2006). It took generations before public opinion turned against racial segregation in the United States, but it cannot be said that the years of debate that preceded that change were in vain. Similarly, the postwar generations in Europe have clung for the past decades to a generous social model that entitled them, among other privileges, to the right to retire at age fifty-five (for some), even in the face of increasing life expectancy and rising budget deficits. It took many years of acrimonious debates (and, to be fair, a massive financial crisis) for the warnings and reasons of deficit hawks to be heard, but it seems that the message is finally sinking in and some reforms are at least being considered.

It thus could be the case that the epistemic properties of deliberation are real and yet too slow to come to fruition for a number of decisions that need to be made under time constraints. The advantage of deliberation, even when it does not yield a decision, is that it renders public the relevant information and arguments, helps frame the problem in terms acceptable to all, and narrows down and structures appropriately the options between which a vote should take place. Let me now propose in what follows a more ambitious and theoretical reply to critics of democratic deliberation, using insights from evolutionary psychology to account for the empirical success and occasional failures of deliberation.

2. A Reply from Psychology: The Argumentative Theory of Reasoning

In this section, I propose that we look at the apparently dispiriting or at least ambiguous empirical performance of deliberation mentioned at the beginning of this chapter through the lens of a new psychological theory of reasoning called "the argumentative theory of reasoning."[5] The motivation for doing so is the hope that the insights of this new theory, whose object is close enough to mine, will bring new arguments, if not evidence, in favor of the thesis that deliberation among the many has epistemic properties. I will show that this new psychological theory not only makes more sense of the available contradictory and ambiguous empirical evidence regarding the epistemic performance of deliberation but justifies a reasonable optimism, particularly against critics who argue that it does not change minds or only changes them for the worse.

The first subsection below presents the argumentative theory of reasoning. The second subsection uses the theory to support the normative ideal of democratic deliberation against Goodin and Niemeyer's criticisms and their defense of "deliberation within" as a better alternative. The third subsection turns to Sunstein's objection based on his hypothesized "law of group polarization" and shows that the argumentative theory of reasoning can make sense of it in ways that preserve my claim for the epistemic benefits of deliberation.

2.1. The Argumentative Theory of Reasoning

The argumentative theory of reasoning (Mercier and Landemore 2012; Mercier and Sperber 2011a, b; Sperber 2001; see also Billig 1996; Gibbard 1990) defines reasoning as a specific cognitive mechanism that aims at finding and evaluating reasons, *so that individuals can convince other people and evaluate their arguments.*

This definition may seem quite intuitive and obvious, but it is in fact a marked break from another theory of reasoning that currently still dominates modern psychology, in particular the psychology of reasoning and decision making. According to this more classical view of reasoning (Evans and Over 1996; Kahneman 2003; Stanovich 2004), reasoning allows us to improve our epistemic status by correcting our own beliefs and intuitions, building on these foundations to reach knowledge and improve the correctness of our judgments and decisions. This theory, although dominant in psychology, runs into problems at the empirical level, where it has been found incapable of accounting for a wealth of data about the imperfections of individual reasoning.

It is indeed well established in psychology that when reasoning is used internally to generate knowledge and make better decisions, its performance is often disappointing. People have trouble understanding simple

---

[5] What follows is based on a joint work with Hugo Mercier (Mercier and Landemore 2012; Landemore and Mercier 2010). It specifically reproduces passages from Landemore and Mercier 2010.

arguments in abstract, decontextualized form (Evans 2002; Wason 1966; Wason and Brooks 1979). Reasoning often fails to override blatantly wrong intuitions (Denes-Raj and Epstein 1994; Frederick 2005). In some cases, more reasoning can even lead to worse outcomes: it can make us too sure of ourselves (Koriat and Fischhoff 1980), allow us to maintain discredited beliefs (Guenther and Alicke 2008; Ross, Lepper, and Hubbard 1975), and drive us toward poor decisions (Dijksterhuis 2004; Shafir, Simonson, and Tversky 1993; Wilson and Schooler 1991). On the classical approach to reasoning, these empirical findings are profoundly disturbing because it seems that human reasoning is flawed and in need of correction. A major problem for the classical theory, in particular, is the stubborn fact of the "confirmation bias"—the empirically well established tendency of individuals to seek out arguments that support a position they already hold (Nickerson 1998). We can see a lot of parallels here with the problems encountered in many empirical deliberative settings.

As a tool for individual use, reasoning thus does not appear particularly compelling. The argumentative theory of reasoning presented here proposes to abandon the classical theory and replace it with hypotheses generated by a different method, that of evolutionary psychology. Whereas classical psychologists resort to sheer intuition to generate hypotheses, evolutionary psychologists rely on the heuristic value of evolutionary theory (Barkow, Cosmides, and Tooby 1992) to make and test predictions regarding behavior and psychological mechanisms that may otherwise seem puzzling, such as lapses in memory (Klein et al. 2002).[6] Evolutionary psychologists thus seek *functional* explanations—that is, explanations that account for features of a trait by appealing to its function, namely, the production of certain beneficial consequences (see also Elster 2007; Hardin 1980). Such functional explanations are particularly appealing when a psychological mechanism introduces some apparent systematic distortion rather than random error, which could be more simply explained by limited capacity. Reasoning certainly exhibits such systematic biases, most strikingly the already mentioned "confirmation bias."

Evolutionary theory suggests that if individual reasoning is so bad at figuring out the truth when used internally, then figuring out the truth

---

[6] It is possible to test an evolutionary hypothesis by pitting its predictions regarding features of the relevant trait against observations of that trait. In the case of the argumentative theory of reasoning, the most prominent example may be that of the confirmation bias. If one of the goals of reasoning is to convince others, then the confirmation bias can be useful by helping the arguer find arguments and convince her audience. The more fine-grained predictions of the argumentative theory regarding the confirmation bias, as well as its other predictions regarding the features and performance of reasoning, cannot be expounded here (but see Mercier 2011a, 2011b, 2011c, and 2011d, 2011; Mercier and Landemore 2012; Mercier and Sperber, 2011a).

cannot be its main function. Based on ideas introduced by Dan Sperber (2000, 2001), the argumentative theory of reasoning has thus been developed to account for reasoning's features, including its biases, in a way that makes evolutionary sense of them. Instead of assuming, as the classical theory does, that the function of reasoning is to allow lone reasoners to improve their epistemic status through ratiocination, proponents of the new theory hypothesize that the function of reasoning is to find and evaluate reasons, so that individuals can convince other people and evaluate their arguments in dialogic contexts. While linking reasoning and argumentation is an old idea, one of the original aspects of the theory lies in the evolutionary hypothesis that reasoning evolved to allow communication to proceed even where trust is limited: the production of arguments may convince people who would not accept others' claims on trust but who are able to evaluate the validity of an argument.

The argumentative theory of reasoning thus breaks with the classical view, first, in the way it reaches its main hypothesis and, second, in the content of this hypothesis. By so doing, the argumentative approach is able to turn what seemed like vices into virtues. If the goal of reasoning is to convince others, then the confirmation bias is actually useful, since it leads to the identification of information and arguments for the side the individual already favors. Likewise, the fact that people are mostly good at falsifying statements that oppose their views is particularly useful if the goal of reasoning is to convince others.

The relationship of the argumentative function to truth finding is complex and worth emphasizing, especially by contrast with the more straightforward classical theory according to which the function of reasoning in individuals is, simply, to figure out the truth about the world. In the argumentative theory of reasoning, we need to distinguish between the complementary tasks of the argumentative function of reasoning, namely, that of producing arguments for one's beliefs and that of assessing the arguments advanced by others in support of their own beliefs. As far as the production of arguments is concerned, reasoning has (and should have) little concern for the pursuit of objective truth, since its main function is to derive support for beliefs already accepted as true. When individuals want to convince others of a given proposition, they generally do not check that the proposition is true since they already believe it. All they are interested in is finding good arguments that support their proposition and are likely to convince the listener. By contrast, as listeners and receivers of arguments, individuals want to be able to *evaluate* arguments, that is, assess their epistemic soundness, in order to decide whether or not they should accept their conclusions. As far as this evaluative task of reasoning is concerned, it is indirectly concerned with

the truth since individuals want to be able to change their minds when epistemically warranted. However, there is a possible tension between the two tasks: even if reasoning is mostly truth oriented when it comes to evaluating arguments, it is still biased by the function of producing arguments. While individuals spontaneously evaluate the soundness of others' arguments, they generally fail to turn the same critical eye on their own argumentative productions.

One implication of the argumentative theory of reasoning is that the "normal" conditions for reasoning are deliberative and social.[7] Reasoning consists in exchanging arguments with at least another person, whom one is trying to convince and whose views one is trying to assess and possibly falsify. This aspect of the theory makes it particularly congenial with the core elements of the paradigm of "deliberative democracy."

The argumentative theory of reasoning uses the following definition of deliberation: *an activity is deliberative to the extent that reasoning is used to gather and evaluate arguments for and against a given proposition* (Mercier and Landemore 2012). Notice that this definition makes reasoning a centerpiece of deliberation, in agreement with the definition of deliberation used in the previous chapter.

Let me emphasize a few key points of the proposed definition. First, the cognitive activity of reasoning—the usage of "central" as opposed to "peripheral" routes—is crucial. Thus, the content of the utterances being exchanged is not all that matters; the way they are generated is important as well (two actors reciting from memory the scripted text of a deliberation would not be deliberating per se). Second, the definition stresses the necessity of an exchange, or more precisely, a *feedback loop* in the reasoning for at least two points of view. Assuming that two people each hold one point of view, the following chain of events is required for genuine deliberation to take place: person A uses reasoning to make an argument from point of view *a*; person B uses reasoning to examine A's argument from point of view *b*, which is at least partially opposed to point of view *a*; person B then uses reasoning to create an argument that partially or fully opposes the previous argument from the point of view *b*; A uses reasoning to examine B's argument from point of view *a*. Notice that the definition of deliberation presented here allows for the possibility of "internal" as well as "external" deliberation, since it is possible even for a lone reasoner to *find arguments for an opposite point of view*

[7] Here it should be noted that the term "normal" has no normative/moral connotations but simply refers to a set of facts about the conditions in which we claim that reasoning evolved (as in Millikan 1987). The normal conditions for the use of reasoning are, according to the argumentative theory of reasoning, those of deliberation with at least another person, and the abnormal ones those of the solitary mind or nondeliberating groups.

than hers.[8] The definition of deliberation embraced by the argumentative theory of reasoning thus draws a sharp distinction between proper deliberation, which centrally involves the mental activity specified above (reasoning), and conversation or discussion, which may not involve this specific mental activity at all.[9]

In its relation to deliberative democracy, the general implications of the argumentative theory of reasoning are the following:

1. It predicts that even genuinely truth-oriented individuals—that is, participants in the deliberation animated by a sincere desire to figure out the truth (e.g., about what the public good requires, or the implications of a given policy) by contrast with partisans, ideologues, or strategic rhetoricians—will have a hard time fighting their hardwired confirmation bias. What the argumentative theory of reasoning implies, therefore, in a nonnormative and purely prudential way, is that if deliberative democrats care about the epistemic properties of democratic deliberation, they should ensure that deliberation is conducted in such a way as to ensure the neutralization of individual confirmation biases.

2. Reasoning is more likely to yield epistemic benefits for both the individual and the group—that is, to be conducive to true or truer individual and collective beliefs—when it takes place in its normal deliberative context.[10] The idea is that even if the function of reasoning is argumentative—to produce and evaluate arguments—rather than purely epistemic, it should still lead to an improvement in epistemic status, both at the individual and the collective level.

[8]If A internally engages in such an exchange of arguments between the two points of view *a* and *b*, then this person is truly deliberating, in her head, with her internal representation of B's point of view. Notice, importantly, that if A finds arguments supporting her own point of view only, then she will still be reasoning, but deliberation will not have taken place. Similarly, a group of people who all think like A and find arguments supporting the point of view *a* are not properly deliberating, even if these arguments are different from theirs, as long as they support the same position.

[9]Manin has himself called "debate" the type of argumentative deliberation that Hugo Mercier and I embrace, by contrast with discussion. We do not use that term but simply narrow down our concept of deliberation to exclude mere discussion from it. Notice also that we are not saying that in the phenomenon of group polarization, people do not reason, simply that they do not deliberate.

[10]Of course, the claims made here are only probabilistic. It is possible for an individual to apply herself and successfully think up counterarguments to her prior views. Similarly, it is possible for like-minded groups to do the same thing, for example, by assigning someone the role of the devil's advocate. The theory simply posits that given what we know of the strength of the confirmation bias and the mental discipline it requires for people to fight it, solitary reasoning or reasoning with like-minded people is less likely to lead to good epistemic outcomes.

Otherwise there would be no point in listening to other people's arguments—and then no point in making any argument.

Before proceeding any further, two potential sources of misunderstanding must be dispelled. First, the claims made by the argumentative theory of reasoning are descriptive—that is, they can be true or false—but they have no implications in terms of what is morally right or wrong or in terms of the norms that ought to guide deliberators. In fact, the argumentative theory of reasoning is compatible with many normative views of politics and does not carry a normative agenda by itself. In particular, even though the argumentative theory of reasoning claims that reasoning has evolved primarily in order to convince others and assess their arguments rather than in order for individuals to figure out for themselves the truth about the world, it does not mean that the theory supports an approach of politics in terms of power struggle, partisanship, rhetoric manipulation, or any such view.

However, it is also the case that despite its lack of a normative agenda, the argumentative theory of reasoning provides theoretical ammunition for the advocates of deliberative democracy who argue on the basis of the epistemic properties of democratic deliberation. The argumentative theory of reasoning is thus not only compatible with an epistemic, deliberative approach to democracy; its predictions converge with the predictions made in this book.[11]

Second, the argumentative theory of reasoning does not imply that reasoning, because of its primarily argumentative function, has nothing to do with truth seeking. This point has already been touched on above, but it bears emphasis. Truth remains on the horizon of the argumentative theory of reasoning, first, because convincing others is more likely to work if the proposed arguments are sound and, therefore, have a connection to truth. People may occasionally be fooled by rhetoric and well-phrased lies, but it is ultimately easier to convince people when what you say is true and can sustain cool and reflective examination. Convincing others is conceptually distinct from merely persuading them. Persuasion has to do with the package in which a message comes (the rhetoric of it), whereas conviction has to do with the substance of the argument.

---

[11] Habermas, one of the most prominent contemporary deliberative democrats, has himself recently stressed the compatibility of detranscendentalized philosophical thinking about the rules that ought to govern the use of our freedom in politics and elsewhere and the "right way to naturalize the mind" in a nonreductionist naturalistic program seeking to identify the causal determinisms regulating human behavior (Habermas 2008: 152–53). The argumentative theory of reasoning can easily be inserted into the program of deliberative democracy, in the same way that Habermas recommends reconciling Darwin and Kant.

Further, the function of reasoning is not just to convince others but also to evaluate their arguments. In this evaluative dimension, truth is also aimed at, either in the recognition of the force of the opponent's argument or, less directly, through attempts at falsifying it. There is thus no necessary disconnection between truth seeking and the function of reasoning, although the relationship is, the theory argues, first and foremost through the function of arguing with others.

To sum up, according to the new theory presented here, reasoning is an argumentative device. Its function is social: to find and evaluate arguments in a dialogic context.[12] The key aspect of the argumentative theory of reasoning is that contrary to traditional classical models, which see reasoning as best deployed in the solitary confinement of one person's mind, reasoning is here assumed to perform best when deployed to argue with other human beings. In other words, reasoning is supposed to yield an epistemic betterment of individual beliefs through the social route of argumentation rather than the individual route of private ratiocination.

How is all of this relevant for the claim that democratic deliberation has epistemic properties? Some deliberative democrats have argued that democratic deliberation does not do much to change people's minds and that deliberative democrats should shift focus from interpersonal exchanges toward "deliberation within." If the argumentative theory of reasoning is true, however, this is likely to be ill advised. The next section considers the case on which Goodin and Niemeyer build their recommendation and shows that it is not strong enough to threaten the prediction of the argumentative theory of reasoning that deliberation with others is a safer bet toward epistemic improvement than "deliberation within."

2.2 Deliberation Within versus Deliberation with Others

Goodin and Niemeyer deplore that since the deliberative turn of the 1990s, most deliberative democrats have moved away from the monological ideal of deliberation at the heart of the early Rawls model of the original position and embraced instead Habermas's later emphasis on actual, interpersonal engagements. For Goodin and Niemeyer, this move away from hypothetical imagined discourse toward actual deliberation is misguided (Goodin and Niemeyer 2003; Goodin 2008[13]) because, for them, it is "deliberation within" rather than talking with

---

[12] Note that saying that the function of reasoning is social does not mean that reasoning has been selected to serve the interests of groups rather than individuals. In fact, reasoning with others directly benefits individuals and only indirectly benefits the group.

[13] While Goodin and Niemeyer's original article is from 2003, all the citations will be to its latest version as chapter 3 in Goodin 2008.

others—or "external deliberation"—that should be the focus of theories of deliberative democracy.

Goodin first coined the expression "deliberation within" as a way to capture the pondering of reasons that goes on in an individual's mind prior to and also during his engagement in deliberation with others. This pondering of reasons involves an exercise in reflection and imagination, in which one is supposed to put oneself in other people's shoes and imagine what their arguments might be. In that sense, deliberation within is not unlike the hypothetical, monological type of ratiocination defended by Rawls. By contrast, Goodin labels "external deliberation" the type of discursive exchanges by which a group collectively ponders the reasons defended by different individuals.[14]

One of the main motivations for this embrace of deliberation within over deliberation with others is, importantly, the unfeasibility of group deliberation on the mass scale of existing democracies and, conversely, the obvious feasibility, at any scale, of deliberation within. If it can be shown that, even in the mini-publics[15] studied by Goodin and Niemeyer, what does most of the work is actually a form of internal ratiocination rather than discursive exchanges with others, there would be reasons to think that where group deliberation is not feasible in the larger public sphere, it can easily be replaced by internal deliberation within each citizen. Thus, particularly on a large scale, the ideal of deliberation would have to be less discursive and interpersonal, and more self-reflective and intrapersonal.

The other main reason why Goodin and Niemeyer think we should return our focus to hypothetical/monological rather than actual/dialogical deliberation is that as far as changing people's minds, deliberation within is also where the action is. Asking "When does deliberation begin?" they answer that not only does deliberation begin in the head of individuals prior to their engagement in social and interactive deliberation but that much of the work of deliberation also ends there. Whatever

---

[14] The choice of "external" to characterize the deliberation that goes on in social settings is slightly misleading in that it suggests that in deliberation with others, ideas are processed in the ether, outside of anyone's heads. Of course, the actual processing of arguments is always taking place in someone's head, not in some fictitious "group mind." But the idea expressed by "external deliberation" is that when many individuals deliberate, their parallel individual reasoning takes as an input the output of at least one other person in the group, instead of functioning in autarky and generating all the arguments pros and cons from the inside. To avoid the ambiguity, this book uses the expression "deliberation with others."

[15] A mini-public is "a deliberative forum consisting of 20–500 participants, focused on a particular issue, selected as a representative sample of the public affected by the issue, and convened for a period of time sufficient for participants to form considered opinions and judgments. Examples of mini-publics include deliberative polling, citizen juries, consensus conferences and citizen assemblies" (Warren 2009: 1)

is later externalized or talked out with others does not do as much as what happens earlier in the privacy of people's minds. For Goodin and Niemeyer, there seems to be both a chronological and epistemic priority of deliberation within over deliberation with others. In fact, regarding external deliberation, they argue that the point of it is mostly one of democratic legitimation (Goodin 2008: 39–40). Besides this legitimation function, external deliberation is supposedly not doing a lot. According to them, "much (maybe most) of the work of deliberation occurs well before the formal proceedings [of public deliberative processes]—before the organized 'talking together'—ever begins" (p. 40).

Goodin and Niemeyer's claim relies on the careful case study of a specific mini-public, an Australian citizens' jury convened in January 2000 to discuss policy options for a controversial road, called the Bloomfield Track, running through an Australian rainforest. The issue was, roughly, to decide how to reconcile the problem of community access and environmental concerns for the unique combination of rainforest and coastal reef endangered by the track. Without going too much into the specifics of the citizens' jury organization, what the analysis brings into relief is how most of the attitudinal changes in jury members took place prior to actual formal deliberation with other jury members, during what Goodin and Niemeyer characterize as the information phase. During this phase, jurors visited the rainforest and the Bloomfield Track and were given background briefings and presentations by witnesses, whom they could then interrogate. This information phase allowed for verbal exchanges between jurors, on site and over tea and lunch at different points, but none of those were organized as the official deliberation phase. Goodin and Niemeyer explicitly define "deliberation" in the narrower sense of "*collectively organized* conversations among a group of coequals aiming at reacting (or moving towards) some joint view on some issues of common concern" (Goodin 2008: 48; my emphasis).

Substantively speaking, what happened during the information phase is that jurors initially concerned about the impact of the Bloomfield Track on the coral reefs nearby were no longer so worried halfway through it. Similarly, jurors who initially worried about the importance of the track for tourism and as an access road for people living in remote northern towns were largely reassured. By contrast, during the discussion phase, a similarly large change occurred in attitudes toward only one proposition. The proposition was "I will be made worse off by any decision about the Bloomfield Track'. While the jurors worried that this might be true throughout the information phase, their fear started to dissipate over the course of the formal deliberation.

According to Goodin and Niemeyer's reading of the experiment, the information phase was much more important than the deliberation phase in transforming jurors' policy preferences:[16] "the simple process of jurors seeing the site for themselves, focusing their minds on the issues, and listening to what experts had to say did all the work in changing jurors' attitudes. Talking among themselves, as a jury, did virtually none of it" (Goodin 2008: 58–59). This would seem to establish the crucial importance of deliberation within and the lesser importance of external deliberation.

Such a finding, they argue, has potentially important implications for deliberative democracy. To the extent that it is possible to extrapolate from the micro-deliberation of a jury to the macro-deliberations of mass democracy, there are lessons to be drawn from the first to improve the practice of the second. Goodin and Niemeyer thus invite us to speculate that much of the change of opinions that occurs in mass democracy could be due not so much to any formal, organized group discussions—presumably those in national assemblies between representatives as well as those taking place in town-hall meetings or during such events AmericaSpeaks or Deliberation Day—but to the internal reflection individually conducted ahead of those, "within individuals themselves or in informal interactions, well in advance of any formal, organized group discussion" (Goodin 2008: 59). This is a rather good thing, in their view, if actual mass-scale group deliberation is unfeasible.

Goodin and Niemeyer find further support for the hypothesis that most of the work is done prior to group discussion in the large literature on persuasion and attitude change, in particular the literature on central versus peripheral routes to the formation of attitudes. Peripheral routes consist in relying on cognitive shortcuts and intuitions to, say, evaluate a candidate's competence—for instance, by relying on party affiliation or attractiveness. By contrast, central routes involve the kind of cognitive effort characteristic of Goodin's internal-reflective deliberative ideal, such as the careful weighing of different positions and arguments—in other words reasoning. Goodin and Niemeyer add that it is important for their argument in favor of deliberation within that "there is nothing intrinsic to the 'central' route that requires group deliberation. Research in this area stresses instead the importance simply of 'sufficient impetus' for engaging in deliberation, such as when an individual is stimulated by personal involvement in the issue" (Goodin 2008: 60). According

---

[16]Two questionnaires measured attitudinal variations on given propositions, such as "Upgrade the track to a dirt road suitable for two-wheel drive vehicles" or "Close the track and rehabilitate the area," as well as subjective assessments by the jury members of what caused them to change their minds. Answers to these questionnaires indicate that what did the most to change individuals' prior beliefs and stabilize their ultimate judgments was the earlier exposure to relevant information and the internal reflections prompted by it, rather than the later discursive exchanges of the "deliberative phase" (Goodin 2008: 49).

to them, the deliberations of this particular Australian jury verify the hypotheses of the model of deliberation within, in that deliberation of the more internal kind did more to change people's attitudes than formal group deliberation of the more discursive sort.[17]

Goodin and Niemeyer's argument runs directly counter to the predictions of other deliberative democrats and the proponents of the argumentative theory of reasoning that deliberation is more likely to have positive epistemic properties in the dialogical context of group reasoning than in the monological context of individual reasoning. Their argument, however, relies on a relatively weak empirical case.

Let me just emphasize three points. Even if one case study was enough to support the case for deliberation within against external deliberation, it is not clear that the example shows as much as Goodin and Niemeyer say it does. A first problem is that the informational phase is far from pure and contains, in fact, a lot of deliberative aspects that could be credited for the change of jurors' minds, rather than any internal deliberation. Goodin and Niemeyer themselves grant that much of the work of the first (informative) phase was done discursively.[18] Insisting, then, on keeping the information and deliberation phases separate and attributing all the merit of opinion change to the first phase, seems artificial. Sure, the "formal official task of the citizens' jury" (Goodin 2008: 48) was in one case gathering information and, in the other, deliberating. But the fact that much of the reasoning that prompted the jury's changes of minds occurred in the first phase cannot be attributed to internal deliberation only, to the extent that it is highly likely that the informal and formal exchanges between jurors at that point contained arguments for or against keeping the track or closing it. Since none of the content of the exchanges during that first information phase is documented—only the content of opinions at different points in the experiment is measured—it is very difficult to judge whether this information phase should not rather be recast as informal deliberation.[19]

[17]They conclude that the deliberative phase "was of much less consequence than the 'information phase'—contrary to the expectations of discursive democrats who would have us privilege conversation over cogitation as politically the most important mode of deliberation" (Goodin 2008: 49–50).
[18]"Witnesses talked, they were interrogated, and so on. There was also much talking among jurors themselves, both informally (over lunch or tea) and formally (in deciding what questions to ask of witnesses)" (Goodin 2008: 47).
[19]To be fair, Goodin and Niemeyer take that objection into account when they remark that the fact that some discussion took place in the first phase of the jury discussion might make it "a model of deliberation in the public sphere of 'civil society.'" In other words, they admit that the experiment did not so much juxtapose an information phase and a deliberation phase as it did two deliberative phases: one informal, one formal. We think this is a rather powerful objection. Goodin and Niemeyer, however, simply counter it by claiming

The second weakness in Goodin and Niemeyer's case is that the fact that jurors themselves perceived that their preferences had changed more during the information rather than the discussion phase[20] could just be an artifact of the way the experiment was presented to them. Even if informal deliberation took place during the first phase, jurors were not allowed by the questionnaire to conceptualize this first phase as discussion but instead as information.

Finally, regarding the relation between the use of central routes and collective deliberation, Goodin and Niemeyer are right to say that "nothing intrinsic" to the central route or "deeper deliberative reflection" requires group deliberation. Yet several times they come close to admitting that group deliberation remains the main and most common impetus for both.

These limitations, partially acknowledged by Goodin and Niemeyer themselves, cast some doubt as to the general validity of their conclusion. As a result, it can be argued that Goodin and Niemeyer's results do not make a strong case for deliberation within as preferable to deliberation with others, at least where deliberation with others is feasible. Meanwhile, the argumentative theory of reasoning remains unharmed in its prediction regarding the epistemic superiority of deliberation with others over deliberation within.

The claim put forward by the argumentative theory of reasoning that the normal condition of reasoning is deliberation with others rather than

---

that the Bloomfield Track had long been a contentious issue within the public sphere of which jurors were already part prior to engaging in that particular jury, so that "something in that initial phase of the jury must have made a difference to them, that informal discussions in the public sphere had previously not" (Goodin 2008: 52). A critique may well grant the point and yet deny that that "something" had anything to do with deliberation within and all to do with a higher motivation to listen to what is said in the mini-public sphere of the jury than to what was ever said in the larger public sphere. I think that the smaller setting of a mini-public might in effect be more conducive to the use of central routes than either the too-large setting of a wide public sphere or the purely internal one of deliberation within. The higher motivation itself could be explained by, say, a heightened sense of efficacy in the smaller rather than the larger public sphere. The point of this remark is just to emphasize that one can come up with an alternative account of the "something" responsible for making the information phase conducive to policy-preference changes that does not require Goodin and Niemeyer's hypothesis of the role of deliberation within. The same analysis about the role of motivation as a stimulant to reasoning could be applied to a recent paper by Muhlberger and Weber (2006), which seems to support Goodin and Niemeyer's conclusion in establishing the superiority of information over deliberation. In both reported experiments, all participants are anticipating the prospect of group deliberation, even if they have not yet taken part in group deliberation or will not formally do so. In both cases, it could very well be this motivating factor that does the work, rather than deliberation within.

[20]Three quarters of the jurors thought discussion was the least important factor in explaining their change of mind (Goodin 2008: 51).

deliberation within does not mean that people can never properly reason by themselves—that is, as Goodin and Niemeyer rightly suggest, make use of central routes on their own. Recent results, in fact, show that it is both possible and sometimes desirable to stimulate deliberation within where it is anticipated that an individual might have to participate in group deliberation (Muhlberger and Weber 2006; see also the literature on accountability reviewed, for instance, in Lerner and Tetlock 1999) and defend their arguments to others. Conversely, it is likely that after experiencing deliberation with others in a mini-public on one subject, individuals are then more capable of replicating the process of reasoning using different perspectives on their own. It is perhaps even possible to "prime" individuals to reason alone in an argumentative sense, despite a natural tendency for them not to do so.

Nonetheless, the argumentative theory of reasoning does suggest a certain priority, if not superiority, of deliberation with others over internal deliberation. In other words, while human beings do not need an actual collective deliberation to be able to reason properly, the fact that the normal conditions of reasoning are those where one naturally encounters a variety of points of view makes it more likely that individuals will use central routes when discussing with others than when reasoning alone.

Talking things out with others, however, is not foolproof, either, and group deliberation will not always have the hoped-for epistemic properties. Let us now see how the argumentative theory of reasoning helps us answer another classical objection raised against democratic deliberation, namely the so-called law of group polarization. The next section explains what the challenge is and how the theory is equipped to answer it.

2.3 The Law of Group Polarization

In different books and in an influential article, Cass Sunstein has argued that a major problem for democratic deliberation and, more generally, deliberative democracy, is what he calls "the law of group polarization" (Sunstein 2002),[21] or the tendency for a group already sharing some views to become more extreme in these views following joint discussion. Of course, there might be some polarizing groups that actually converge on the truth, so polarization need not always indicate that deliberation makes things worse rather than better. If polarization is a law that applies no matter what the original consensus, however, it is highly dubious

---

[21] The other problems affecting deliberation are, according to Sunstein, the fact that human deliberators are subject to heuristics and biases, the existence of hidden profiles and the common-knowledge effect, and the possibility of informational and reputational cascades.

that in most cases the polarization effect is connected to any epistemic improvement. In fact, even occasional convergence toward the truth—an undeniable epistemic improvement—might be achieved accidentally and thus fail to provide an argument for deliberation. As Sunstein suggests, if group polarization is such a systematic phenomenon, it would seem to provide a strong argument for turning away from democratic deliberation, even where it is feasible, toward a mere aggregation of individual judgments, individual deliberation within, or a combination of both. Furthermore, this alleged law apparently contradicts our prediction that reasoning with others improves the epistemic status of individuals (and indirectly, of the group as well).

According to Sunstein, the law of polarization accounts for why after discussion, a group of moderately pro-feminist women will become more strongly pro-feminist, for example, or why citizens of France will become more critical of the United States and its intentions with respect to economic aid, or why whites predisposed to show racial prejudices will offer more negative responses to the question of whether white racism is responsible for conditions faced by African Americans in American cities.[22] In order to explain group polarization in such cases, Sunstein turns to two well-established (theoretically and empirically) mechanisms, already encountered at the beginning of this chapter. The first involves social influences, that is, the fact that people want to be perceived favorably by other members of the group. Such tendencies create a pressure to conform to the perceived dominant norm. The result is to move the group's position toward one or another extreme, and also to induce shifts in individual members.

The other mechanism is the limited pool of persuasive arguments to which members of the group are exposed, and the path dependency this creates toward more extreme versions of foregone conclusions. To the extent that individuals' positions are partly a function of whichever convincing arguments they are exposed to, and to the extent that a group already prejudiced in one direction will produce a much greater amount of argument for one side of the case than for the other, group discussion can only reinforce individuals' prior beliefs. In Sunstein's words, "the key is the existence of a limited argument pool, one that is skewed (speaking purely descriptively) in a particular direction" (Sunstein 2002: 159).

Deliberation as celebrated by deliberative democrats, Sunstein concludes, is overrated. In his view, the underlying mechanisms of group deliberation "do not provide much reason for confidence" (Sunstein 2002: 187). He even suggests that since we do not have any reason to think that

[22] All examples are from Sunstein 2002: 178.

deliberation is "making things better rather than worse," and given the prevalent mechanisms behind the law of group polarization, "the results of deliberative judgments may be far worse than the results of simply taking the median of pre-deliberation judgments" (p. 187). As is the case for Goodin and Niemeyer, the value of democratic deliberation for Sunstein is more procedural than epistemic. In fact, whereas Goodin and Niemeyer only suggest the relative neutrality of democratic deliberation (it did not change people's views much, but the little it did was not necessarily for the worse), Sunstein thinks that deliberation changes people minds by polarizing them, and most likely for the worse. Are Sunstein's warnings about the risk of group polarization a reason to give up on deliberation with others and embrace deliberation within or an aggregation of views based on those internal deliberations instead? Should we, in other words, give up on the ideal of deliberative democracy? I think not.

First, the problem of group polarization does not constitute a reason to embrace deliberation within over external deliberation, because even if the pressure to conform did not affect the lone reasoner (which nothing guarantees), the limited pool of arguments certainly does. Second, the case against deliberation with others that one may be tempted to build on Sunstein's law of group polarization suffers from the fact that this law of polarization really only applies to a type of communication that fails the standard of deliberation as both the proponents of the argumentative theory of reasoning and most deliberative democrats define it.

Recall that in the definition given in section 1, deliberation must involve a genuine consideration of arguments for and against something. An interpersonal exchange in which arguments for both sides are not properly considered does not count as deliberative. From that point of view, many of the discursive exchanges among like-minded people described by Sunstein—groups of feminists, anti-American French, or racist Americans—are likely to fall short of the requirement of deliberation as defined in this book. The fact that such exchanges lead to polarization is therefore not an indictment of deliberation properly construed but of something else, which may look like it but really isn't and at best deserves the name of "discussion."[23] The key for deliberation is not just having diverse arguments but having arguments that respond to each other in critical, even conflicting ways.

---

[23] Manin similarly points out that the main difference between groups that tend to polarize, like those observed by Sunstein, and groups that tend not to, like the citizens' jury studied by Goodin and Niemeyer or the members of James Fishkin's deliberative polls, is that the former groups did not seem to properly take conflicting views into consideration. Manin further insists that, contrary to what many authors besides Sunstein emphasize (e.g., Bohman 2007: 348–55), diversity of views is not enough, since even people with different perspectives may fail to engage each other's arguments in the kind of adversarial manner conducive to epistemically satisfying deliberation (Manin 2005: 9).

Sunstein calls exchanges among like-minded people "enclave delibera-tion" (Sunstein 2002: 177, presumably after Mansbridge 1994). Yet, even among like-minded people, there is a difference between an argumentative exchange, that is, an exchange that genuinely pits arguments against each other, and an exchange of diverse and yet self-reinforcing views. Sunstein's focus on the fact that people in the polarizing groups of his examples share the same initial views is slightly misleading in that it is not just (or even) the starting point of the deliberation that counts as most important but the fact that the exchanges are truly deliberative and based on arguments that oppose each other. The same way that individuals can make an effort to consider different viewpoints internally, like-minded groups should be able to take into account perspectives beyond those represented in the group. As with individuals, not all like-minded groups are bound to polarize. The advantage of a group over an individual, however, is that all things being equal otherwise,[24] the greater the number of people in the group, the less likely they are to be all perfectly like-minded, hence increasing the chances that a genuinely conflicting perspective can trigger genuine deliberation.

Sunstein briefly raises the possibility that the kind of group exchanges among like-minded people that he considers—"enclave deliberation"—does not qualify as proper deliberation.[25] To this objection, his terse reply is that "[i]f deliberation requires a measure of disagreement, this is a seri-ous question" (Sunstein 2002: 186). He nonetheless goes on to argue that "even like minded people will have different perspectives and views, so that a group of people who tend to like affirmative action, or to fear global warming, will produce *some kind of exchange of opinion*. I will urge that in spite of this point, enclave deliberation raises serious dif-ficulties for the participants and possibly for society as a whole" (Sun-stein 2002: 186; my emphasis). At this point in the paper, and regardless of whatever problems arise from group polarization for individuals and society, Sunstein has in fact conceded the main issue: discussion among like-minded people who fail to consider pro and con arguments for a given position produces at best "some kind of exchange of opinions." This kind of exchange should not count as deliberation. Furthermore, the law of polarization, which turns out to apply only to groups of like-minded people who do not properly deliberate, can no longer be used as an argument against democratic deliberation.

---

[24] "All things being equal otherwise" means here that I assume that increasing the num-ber of people does not concomitantly introduce or increase value diversity about the ends to be achieved. We need a diversity of perspectives that does not also introduce a diversity of values and notions of the goals to be achieved by the group. As stated in the introduction and elsewhere, this book is built on the premise that this constraint can be satisfied.

[25] Interestingly, the question—"Should enclave deliberation count as deliberation at all?" (Sunstein 2002: 186)—was not even addressed in the earlier unpublished version of the paper, available at http://www.law.uchicago.edu/files/files/91.CRS.Polarization.pdf.

If we now consider only the cases where deliberation as an exchange of arguments for and against something, rather than as any form of discussion or exchange of views, is at play, the results are much more positive. As emphasized in the definition given earlier, several conditions have to be met for a discursive exchange to count as deliberative. In particular, arguments between several opposing points of views have to be debated. When this is the case, deliberation does tend to produce good reasoning, which in turn produces good outcomes, in terms of improving beliefs and related conclusions (see Mercier and Landemore 2012 for a review). For instance, when people have different and conflicting opinions on questions about which there exists a factual answer, deliberation improves performances, sometimes dramatically (e.g., Sniezek and Henry 1989). This also applies when there is no strictly superior answer but one can distinguish between better and worse arguments offered in support of a given alternative (e.g., Laughlin, Bonner, and Miner 2002). The good performance of reasoning in the context of genuine deliberation is also supported by a large quantity of studies on teamwork in the workplace and at school (e.g., Michaelsen, Watson, and Black 1989; Slavin 1996). Finally, evidence of good performance has been found in studies of deliberating citizens, in the context of deliberative polls, citizens' assemblies, citizens' juries, consensus conferences, and even hybrid forums that mix both regular citizens and experts (e.g., Fishkin 2009; Warren and Pearse 2008). In the case of citizens' assemblies or consensus conferences, for example, there is no right or wrong answer identifiable or knowable *ex ante,* but one can use as a proxy the general consensus of observers of those experiments, whether professional politicians, the larger public, or experts brought in as impartial observers. The general consensus is that the deliberating groups of citizens in these cases ended up with more-informed beliefs and, where relevant, compelling policy proposals.[26] All in all, these results seem to indicate that in politics as in other areas, group reasoning—that is, individual reasoning practiced in the context of an argument with others—yields very good results, which generally surpass those produced by the solitary mind.

Not all groups of like-minded people are doomed to polarize, provided they contain at least some dissenting individuals and make an effort to take their arguments seriously—that is, make an effort to reason as opposed to letting their confirmation bias run unchecked and produce only arguments for the side they already favor. In other words, it is not always a problem that a group of people who are about to deliberate

[26] Gerry Mackie provides an important methodological caveat for these studies. He notes the effects of deliberation are "typically latent, indirect, delayed, or disguised" (Mackie 2006: 279), and that therefore some studies may fail to observe them even though they are real. This argument therefore strengthens any positive results actually obtained.

starts with a lot of shared views. After, all those views might be the right ones! The problem is when people share the same opinions and make no effort to test them against opposing views.

If groups of like-minded people might not necessarily polarize, conversely, diversifying the pool of arguments might not be enough to prevent polarization. Playing with the thought experiment of a deliberating body consisting of all citizens in the relevant group, whether a community, a nation, a state, or the whole world, as opposed to just a subset of them, Sunstein argues that the main advantage of such a setting is "by hypothesis, the argument pool would be very large" (Sunstein 2002: 189). While Sunstein does not think that such an all-inclusive deliberating body would be ideal—social influences may remain, and biases too—his suggestion is that at least it would "remove some of the distortions in the group polarization experiments, where generally like-minded people, not exposed to others, shift in large part because of that limited exposure." Sunstein then concludes that full information, by which he means knowledge of all the relevant facts, values, options, and arguments that may affect a decision, is the best antidote to polarization (pp. 191–92).

As Sunstein acknowledges, though, full information is a daunting requirement. In practice, not everybody can have all the necessary information before starting to deliberate. Moreover, not only is full information a daunting requirement, but it is not a sufficient condition to remove the risk of group polarization.[27] Contra Sunstein's claim that "if there is already full information, the point of deliberation is greatly reduced" (Sunstein 2002: 188), it can be argued that even with full information there is still a need to talk things out with others.

Another way to say this is that what is needed is not just diversity of inputs, or informational diversity, but an actual deliberation built on this informational diversity and, in effect, a diversity of reasoning processes. In other words, to be truly effective, informational diversity must be used in deliberative exchanges—exchanges in which one piece of information or argument is taken as the starting point of another argument for or against it. Depolarization can occur even if the pool of arguments and information is far from exhaustive.[28] In other words, it is not enough, and it is not perhaps even necessary to ensure that people in the group

[27] In order to thwart the limited-pool-of-arguments mechanism, Sunstein indeed suggests that "[p]erhaps group polarization could be reduced or even eliminated if we emphasized that good deliberation has full information as a precondition; by hypothesis, argument pools would not be limited if all information were available" (Sunstein 2002: 188).

[28] Of course, if information—the argument pool—is too limited and too biased, deliberation based on it won't do miracles. But between the extremes of seriously limited and biased information on the one hand and full information on the other, there is a space in which deliberation can have transformative and epistemic properties, even among initially like-minded people.

hold different views from the get-go. One must more crucially ensure that they engage in the sort of deliberative exchange that can even produce correct arguments when those are missing or go unacknowledged. In other words, one should focus on both the inputs and the process of deliberation rather than on its input only.

According to the argumentative theory, the participants' biases are likely to play a much more important role than the initial distribution of information. Even full information cannot guarantee an epistemically sound outcome if everybody agrees on the issue to start with: each individual's confirmation bias is likely to make for a biased discussion in any case. The confirmation bias can, however, be put to good use. When group members disagree, they will still be more likely to find arguments for their own side of the issue, but the consequences of their bias can then be positive: it guarantees a more exhaustive exposure to the arguments supporting the different sides of an issue. As a bonus, there is no need for full information prior to the debate: full (or at least, more complete) information is precisely one of the main achievements of the debate. In such a case, far from being a nuisance, the confirmation bias becomes a form of division of cognitive labor.

Sunstein's argument thus blows out of proportion an epistemic failure that only affects groups of like-minded people that do not engage in deliberation per se, and in so doing, he fails to see that there is something more fundamentally problematic than social homogeneity. While it is true that "group polarization helps to cast new light on the old idea that social homogeneity can be quite damaging to good deliberation" (Sunstein 2002: 177), the real question is thus: why do socially homogenous people fail to consider other arguments, and how can we institutionalize collective decision making so as to remedy this phenomenon? The point is that in the end, it would not matter much if a deliberating group were made up only of lawyers or men or white people or well-to-do people if they were not also more likely to argue along the same lines or for the same things.[29]

The theory of reasoning as arguing predicts that only discursive exchanges among like-minded people that do not involve a proper weighing of the pros and cons will lead to polarization, but not necessarily discussion among people who start with the same information and argument pool but engage in proper deliberation, even on a skewed informational basis.

[29] The reason to include all the members in the deliberative body in the thought experiment proposed by Sunstein is that we are more likely to obtain arguments opposing the dominant view this way than through any other democratic means. Unlike social heterogeneity, indeed, opposition of points of views and arguments is less easy to identify prior to actual deliberation, so it becomes difficult to a priori select the "right" social mix that would ensure the relevant opposition of arguments.

CONCLUSION

The argumentative theory of reasoning used in this chapter yields predictions regarding the epistemic properties of deliberation that support the claims made in the previous chapter. The argumentative theory of reasoning allows us to predict where deliberation is likely to work well (in contexts that fulfill or approximate the normal circumstances for which it was designed, i.e., deliberative contexts) and when it is likely not to (in abnormal contexts, i.e., solitary reasoning or certain types of collective exchange among like-minded people).

Most important for the argument of this book, if the theory of reasoning as having primarily a social function of conviction and evaluation of other people's claims is correct, combined with assumptions about the way confirmation biases tend to cancel out in diverse groups, it lends plausibility to the claim implicit in the normative ideal of many deliberative democrats and the explicit claim put forward by this book that deliberation with others has epistemic virtues and, additionally, more epistemic virtues than reasoning on one's own.

The argumentative theory of reasoning also suggests that there is nothing utopian about the demands placed by deliberative democrats on individual reasoning. In its normal, dialogical context, reasoning will do well what it is supposed to do; confirmation biases are actually harnessed to epistemic benefits. There is no need to deplore or try to fix the limitations of individual reasoning. What matters is to set up the optimal conditions for it: genuine deliberation with others.

Let me end this chapter by briefly tying the conclusions reached in this chapter back to the idea of cognitive diversity. There may indeed appear to be a tension between the claims of this chapter and the previous one. Chapter 4 upheld the claim that "more is smarter" due to the statistical probability that "more" means greater cognitive diversity. Yet in this chapter we saw that it is fine to have a smaller set of like-minded individuals deliberate, so long as they deliberate genuinely—that is, they really consider alternative arguments and weigh the pros and cons of each. The tension is only apparent, however, and readily solvable. First, I should reiterate that the goals of these two chapters are sensibly different. The previous chapter meant to establish the epistemic properties of deliberation to the extent that it is properly democratic, that is, inclusive of all the members in the group. This chapter, by contrast, is primarily concerned with defending the epistemic properties of deliberation in general, regardless of whether it is inclusive of a lot of people or not. Second, while it is true that a smaller group that deliberates properly can be as smart as a larger one, it remains simply more likely on average that better deliberation will take place where diversity already exists, as a natural by-product of a large number of people being involved, rather than where deliberators need to be disciplined to be able to "simulate" diversity by considering other arguments than the one they spontaneously think of. Even such "simulations," in fact, may require something like cognitive diversity, as different people may think of different alternative arguments, and have different approaches to them.

CHAPTER SIX

Second Mechanism of Democratic Reason: Majority Rule

SINCE DELIBERATION IS NOT A PERFECT or a complete decision mechanism, let us now turn to a very efficient decision mechanism which, though undeniably democratic, is often looked at with great suspicion: majority rule. I will here focus strictly on simple majority rule, defined as the rule by which one of two alternatives is selected, based on which has more than half the votes. Simple majority rule is the binary decision rule used most often in influential decision-making bodies, including the legislatures of democratic nations. This chapter argues that simple majority rule is an essential component of democratic decision making with its own distinct epistemic properties and a certain task specificity, namely a predictive function: majority rule is ideally suited to predict which of two options identified in the deliberative phase is best. This is not to say that prediction is the only function for which majority rule can be used—it can also be used to arbitrate fairly between competing incommensurable preferences—but in an epistemic framework, I argue that prediction is the function of majority rule in the same way as chapter 4 has argued that the function of deliberation is to figure out the solution to collective problems.

This talk of "prediction" is not meant to suggest that majority rule is oriented only toward future events. Yet in many voting situations it will be the case because the truth of the matter is hidden behind the veil of the future, so our prediction is verified at some point after we make it. Think, for example, of the situation in which we are trying to assess the future competence of a candidate running for president. When we vote for him, we are at least in part making the prediction that he will be competent. A few years down the road, it might become obvious whether we were right. In other cases, however, the truth of the matter will remain hidden forever, as in the case of a jury deciding whether a person is guilty, with no ulterior information justifying reopening the case and questioning that judgment. The decision made by the jury in such a case consists in a prediction about whether or not a fact is true, a prediction that we will never be able to verify directly. The existence or inexistence of a verification mechanism is not what determines whether the vote or jury decision counts as a prediction. What determines the predictive nature of the judgment in both cases is whether or not the goal is to figure out the right answer to a given question, where the right answer is defined as the factual or moral "truth" of the matter:[1] Will the candidate be a competent president? Is the defendant guilty or innocent?

As far as the predictive function of majority rule is concerned, there are good reasons to think that majority rule is more epistemically reliable than minority rule. In light of such reasons, one can further argue that majority rule should not only be seen, as it is by most deliberative theorists, merely as a fair way to settle on a collective decision when time is running out for deliberation. It should be seen, also, as a way to turn imperfect individual predictions into accurate collective ones, together with a pre-voting deliberative phase that formulates the choice options, forming an epistemically efficient system for collective decision making.[2]

I will argue that this task specificity of majority rule makes it a supplement to, rather than a rival of, deliberation. Indeed, majority rule is constitutively unable to formulate the options that are voted on. Pitting majority rule against deliberation, aggregative democrats have sometimes argued for a silent, supposedly Rousseauian democracy, one in which voting would take place without "external" deliberation, because of the possible corrupting influence of discussion on individuals' judgments. This type of comparison between majority rule with and without a pre-deliberative phase is misguided, however, in that it forgets that the options on the voting menu cannot be formulated without a deliberative phase. What aggregative democrats implicitly suppose—not so surprisingly, since they are often also Schumpeterian elitist democrats—is that the choice options are handed to the people from above by supposedly more enlightened elites. But the fact is that even for the model of silent voting, some kind of deliberation must have taken place prior to the vote, to decide what options people are going to vote on. In this book, I'm interested in comparing pure models of decision making—that is models where the deliberative and aggregative phases are either oligarchic through and through or democratic through and through—so I will not consider the epistemic properties of mixed combinations, where the deliberation takes place among the

---

[1] See chapter 8 for a defense of the existence of something like "moral facts."

[2] To the extent that majority rule aggregates individual predictions or judgments about the future, it can also be compared with polls and other snapshots of public opinion. Although polls are not actual decision mechanisms, they are frequently used as a basis for political decisions, so the comparison with majority rule has some relevance. Other predictively accurate aggregative decision procedures are "information markets." As their name indicates, what these markets aggregate is information (whereas voting aggregates, as I will argue later, not just information, but preferences and values as well). The accuracy of these markets is well established (see Surowiecki 2004; Servan-Schreiber 2012, and the appendix of this chapter).

few and then the vote takes place among the many or, conversely, deliberation is democratic but followed by an oligarchic vote.

The question of whether, once the options have been democratically defined through a pre-voting, deliberative phase, deliberation should continue about the respective merits of the options presented is another question—one about, perhaps, the diminishing returns of deliberation past a certain point. But assuming that there can be such a thing as a "silent" democracy is absurd. Rather than try to compare the epistemic properties of majority rule with and without deliberation—a comparison I don't find particularly meaningful—I will simply compare majority rule and minority rule assuming that a similar amount of relevant information and arguments have been made available to all prior to voting. I will also consider the comparison between deliberation—to the extent that it produces a unique decision—and majority rule.

This chapter includes three sections, which successively address three distinct accounts of the epistemic properties of judgment aggregation through majority rule: the Condorcet Jury Theorem, the so-called Miracle of Aggregation, and a transposed account of Scott Page and Lu Hong's model of the wisdom of crowds based on cognitive diversity. The latter seems to me a particularly plausible account of why collective predictions are better than individual predictions. Even though I just said that I am not interested in comparing the epistemic properties of deliberation and majority rule per se (at least not such that one is mutually exclusive of the other), it is worth noting that all the accounts covered in this chapter suggest a general epistemic inferiority of sheer judgment aggregation when compared with pure deliberation: unlike deliberation, sheer aggregation of judgments does not allow for the weeding out of bad information and ideas from good ones. Again, I don't think this means we should privilege deliberation over majority rule, since neither procedure is fully independent of the other. It does suggest, however, a chronological priority of deliberation over majority rule: one is better off moving to majority rule when the epistemic properties of deliberation have been exhausted.

Furthermore, what these results suggest is that all things being equal otherwise, the aggregated prediction of a large group of average citizens will not necessarily be more accurate than the aggregated prediction of a small group of experts. This need not lead us to antidemocratic conclusions but invites us to see majority rule as just one cog of democratic reason.

1. The Condorcet Jury Theorem

We already covered the basics of the Condorcet Jury Theorem (CJT) in chapter 3. I will here consider only its plausibility as an account of the epistemic value of majority rule. Recall that the theorem demonstrates that among large electorates voting on some yes-or-no question, majoritarian outcomes are virtually certain to track the "truth," as long as three conditions hold: (1) voters are better than a coin flip at choosing the true proposition; 2) they vote independently of each other; and (3) they vote sincerely, as opposed to strategically. For ten voters, each of which has a .51 probability to be correct on any yes-or-no question, a majority of six individuals will have 52 percent chance of being right.[3] For 1,000 people, a majority of 501 is almost 73 percent sure to be right. As the group grows larger, the majority gets closer and closer to a 100 percent certainty of being right.[4]

Formal theorists today generally acknowledge the merits of the Condorcet Jury Theorem, which has spawned many formal analyses in recent decades. These analyses, however, usually fall short of drawing substantive normative implications.[5] The most common criticism of the theorem is that the assumptions are pure mathematical technicalities with no matching real-life application (e.g., Ladha 1992).[6] Meanwhile, more philosophically oriented theorists remain unsure of the relevance of the CJT for democratic theory. Even David Estlund, an unambiguous advocate of an at least partially epistemic theory of democracy, who repeatedly mentions the CJT in support of the superiority of majoritarian decisions over alternative decision rules, ends up distancing himself from the theorem as "less than trustworthy," and more recently as entirely "irrelevant" (Estlund 1997: 189; 2008: chap. 12). Other philosophers, like Elizabeth Anderson, similarly consider the CJT as a less than convincing account of the epistemic properties of democracy and quickly turn to an alternative, more deliberative account (Anderson 2006). Krishna Ladha summarized the general feeling about the CJT when he commented: "the assumptions preclude any application of the theorem to democratic politics" (Ladha

[3]Notice that a majority of 501 means that at least 501 persons voted a certain way (i.e., probability of 73 percent represents the sum of probabilities that exactly 501, 502, 503, and so forth all the way to 1,000 persons voted the same way). See List 2004b for a discussion of the relevant statistical fact.

[4]See chapter 6 for more details and the appendix to this chapter for graphic illustrations of the law of large numbers.

[5]Typically, formal analyses (e.g., Ladha 1992) shove to the side the question of whether it makes sense to talk about "right" and "wrong" answers in politics and assume it away in order to focus on the mathematical subtleties of the theorem. But this question needs to be addressed in order for the theorem to have any relevance for democratic theory.

[6]Duncan Black set the mood for this reception of the theorem in this respect. Despite a sincere admiration for Condorcet's finding and its value for group-decision theory in particular, Black was skeptical about the wider applicability of the theorem (Black 1958: 185). He was particularly worried about the idea of a "probability of the correctness of a voter's opinion"—a phrase that he considered "without definite meaning" (p. 163).

1992: 617). More recently, Martí (2006: 39) similarly reports that what most critics argue against the CJT is the impossibility of meeting in practice the idealized conditions required by the CJT, namely sincere voting, independent voting, and epistemic average competence above .5, as well as the premise that the choice is between two options. In the following, I try to establish the normative relevance of the theorem by arguing that, granting a certain idealization, the assumptions are more empirically plausible than critics allow.

1.1 Binary Choices

Let me start by defending the CJT against what I take to be the weakest objection, namely the view that it is implausible to assume that political choices are always binary. The CJT does indeed assume a framework for voting in which choices are between two options only: representatives vote for the status quo or a specific amendment to it; citizens vote yes or no to a referendum question or choose between no more than two candidates. Objectors have pointed out that this assumption weakens the relevance of the CJT, since in real life, many choices involve more than two options.

The main advantage of restricting the analysis to situations of binary choices is that one avoids the problems theoretically encountered in the case of three or more options, such as cyclical majorities (Arrow's Impossibility Theorem) or the problem of strategic voting (Gibbard-Satterthwaite theorem). These two theorems have proven extremely damaging, at least in the ways political scientists have interpreted them, for majority rule and democracy in general. Arrow's Impossibility Theorem states that no social choice function can aggregate individual preferences in a way that does not violate one or more of a set of a priori reasonable conditions. It has been used to establish the indeterminacy and meaninglessness of democratic choices (Riker 1982). The Gibbard-Satterthwaite theorem can be used to reach similar conclusions, since it states that majority rule is not immune to strategic voting for three options or more.

I consider in the next chapter some of the attempts that have been made to take the bite out of the theorems by various talented political theorists and scientists. Here I will simply argue that considering only scenarios where two options are at stake is both theoretically defensible and empirically plausible.

Empirically, it is simply a fact that in existing democracies, many assemblies break down their decisions into pairwise votes, with one option supporting the status quo and the other amending it (see also Risse 2004). Theoretically, one can reply to the objector that a vote involving multiple choices—say, candidates A, B, and C—can always be redescribed as a sequence of binary choices: opposing A to B, then whoever wins to C; or opposing A to C, then whoever wins to B; or opposing B to C, then whoever wins to A. Should pre-voting deliberation fail to reduce the choice to two options, the task of determining the order in which pairwise voting between competing options would be organized could be entrusted to an agenda setter.

The problem of reducing a multiple-choice situation to a sequence of binary choices is that if there is a choice between multiple agendas and if the choice of agenda can be expected to influence the outcome, then strategic actors will care about and may try to influence the choice of agenda. Against this risk, one possibility would be to use a random mechanism of agenda setting. Another option could be simply to ensure the political accountability of the agenda setter to the larger community (e.g., through election or peer pressure or some other mechanism) so as to check his or her possible partisan motivations. While some potentially epistemically damaging indeterminacy may result from the order in which the pairs are formed, this indeterminacy seems an unavoidable cost of decision making, whether the decision maker is a minority or a majority. The important point here is simply to point out that a binary framework is perfectly plausible, whether one supposes an agenda setter or a pre-voting deliberative phase that generates a portioning of political problems into binary options.

Admittedly, even bracketing the issue of cyclical preferences, which only arises for three or more options, the defense of majority rule one could build on the CJT potentially runs into other problems, such as the discursive dilemma (also called the doctrinal paradox). I consider this problem in the next chapter.

1.2 The "Enlightenment" Assumption, or Why Trust the People?

Besides the general premise of a binary choice, there are three assumptions under which the CJT operates and which have been historically subject to critique as infeasible. The first one is that the average voter can be trusted to be right more than half the time on any binary political issue. This assumption of general enlightenment raises skepticism, of course. Chapter 2 has sufficiently illustrated the depressing portrait of the average voter painted by contemporary political science: rational in theory, the average voter is in practice ill informed and minimally reasoning. Condorcet himself was well aware of the limits of his enlightenment assumption. Given the "prejudices" and "ignorance" common among people, including about "important matters," Condorcet remarked, "it is clear that it can be dangerous to give a democratic constitution to an unenlightened people. A pure democracy, indeed, would only be appropriate to a people much more enlightened, much freer from prejudices than any of those known to history" (Condorcet 1976: 49). Condorcet initially meant his theorem to apply to small assemblies of superior men, like a representative assembly. Interestingly, the few contemporary commentators who find the CJT plausible as a defense of the epistemic properties of majority rule similarly only consider the case of majority rule used among representatives. Waldron, for example, considers the CJT as a plausible argument to defend "the dignity of legislation" but, following what he takes to be Condorcet's own belief, finds it no longer appealing when it comes to too large a body of decision makers, which would seem to rule out popular referenda (Waldron 1999: 32).

Is it really the case, though, that representatives are more enlightened than their constituents? This Burkean approach to the function of elections as selecting the best and the brightest can be contrasted with the more democratic reading of representation attempted in chapter 4. If the function of representation is simply to reproduce on a smaller scale the diversity of the larger group, then there is no reason to want or expect that representatives are smarter than their constituents. It is interesting to note that over the years, even an eighteenth-century aristocrat like Condorcet came to endorse a more optimistic trust in the reasonability of the people at large and, consequently, universal suffrage.[7] Why is it, then, that contemporary political scientists, who are all by and large democrats, should yet be so skeptical about the capacity for self-rule of their fellow citizens?

Political scientists sometimes seem almost more skeptical than eighteenth-century aristocrats about the enlightenment of regular citizens. Yet the large amounts of data documenting voter incompetence only prove so much. Some have even argued that there is an elitist bias built in the measurement of this incompetence so that the measurement ultimately reflects nothing more than existing preconceptions. Critics have thus offered a rebuttal of many of the pessimistic conclusions built on that evidence (Lupia 2006). Further, even assuming some topical incompetence on specific, more technical issues, one can argue that

---

[7]One reason for his change of mind is perhaps that he found empirical evidence for people's gradual enlightenment during his own lifetime. Baker notices that Condorcet had a gradual evolution with respect to voting rights. He describes Condorcet moving progressively, from the theoretically inconsistent but pragmatic defense of limited suffrage to a perfectly consistent but almost utopian (in most of his contemporaries' view) defense of universal citizenship including the poor, women, and even black people (Baker 1982). At least in the case of women, however, it is unclear that Condorcet was willing to grant them specifically the right to vote, as opposed to a more vague "right of citizenship." For an excellent analysis of Condorcet's ambiguities on that topic, see Fauré 1989

voters' competence is sufficient for a number of "big questions," including general policy orientations that may be the object of a referendum, or
the choice of competent representatives (Lupia 2001). Chapter 7, in that
same spirit, will at any rate attempt to debunk the claim that citizens are
too poorly informed to vote smartly.

At this point, let me just remark that it would be odd to assume that,
in their daily life, human beings' cognitive abilities are below those of
a coin flip. And if we can assume a better-than-random level of competence at the individual level, there is a priori no reason to think that
when it comes to collective decisions about general political issues,
such as whether the country should pursue fiscal responsibility or promote employment in a time of crisis, individuals all of a sudden become
incompetent—pace contrived rational choice theory predictions (which
I also address in the next chapter). I would argue that trust in the public
does not require a leap of faith, but a simple belief in common sense,
which has been said to be the most fairly distributed thing in the world.[8]
If individuals have enough ability to find their own way in life and
develop reasonable moral intuitions about the right thing to do as individuals, why, on the face of it, should it be any different when it comes
to public decisions?

A Hayekian might object that finding one's way in life only requires
a local knowledge readily available to individuals, whereas many if not
most political questions involve a global knowledge that is not available
to anyone in particular, but especially not to the average citizen. The first
answer one may make to this critique is that if it is global knowledge that
is required in order for individuals to be able to vote or have an opinion,
then the criticism overshoots, since professional politicians are no better
situated than average citizens. None of them can claim, as individuals, to
have access to any form of global knowledge. What Hayek has in mind as
an alternative to citizens' local knowledge is not the knowledge of professional politicians but the knowledge contained in the impersonal device
of the omniscient market. As I already stressed in this book, however, the
question of whether to resort to political or market mechanisms is largely
orthogonal to the question of the best *political* rule. As far as this latter
question is concerned, this book actually argues that we are more likely
to obtain some form of "global knowledge" through the combination
of deliberative and aggregative procedures that involve the many rather
than those involving the few only. So I am not denying that some form
of global knowledge is required in politics, but I believe that it can be

---

[8]Descartes's famous claim was not backed up by sophisticated empirical surveys and
measurements, but it has undeniably an empirical flavor, being based on the sample of
people from all walks of life that Descartes himself had encountered in his lifetime.

produced, at the level of the system, by the proper aggregation of individuals' limited and yet pertinent knowledge.

However counterintuitive this thought may have become in contemporary political sciences, I would thus argue that the average citizen cannot be easily presumed to be worse than a coin flip on general political issues.

1.3 Independence

The assumption that individuals vote independently of each other forms, according to many commentators, the main weakness of the Condorcet Jury Theorem. According to some, the problem is that it seems to require that there be no opinion leaders and no communication among voters, not even the sharing of common information, culture, religion, or beliefs, or other elements that could lead to correlated votes (e.g., Ladha 1992: 621). On this very demanding reading of independence as lack of correlation between the votes, the Condorcet theorem becomes wildly implausible. Votes are likely to be correlated because in real life, voters share common information, communicate with each other, and are influenced by various schools of thought or opinion leaders espousing the same or opposite opinions. Empirical studies (e.g., Panning 1986) show that individual judgments are seldom independent because of mutual influence and communication.

The CJT has, however, been generalized to the case of correlated votes. As long as the average interdependence between votes is sufficiently low, the theorem holds (Ladha 1992: 626). Relaxing the independence assumption into low vote correlation permits pre-voting communication and deliberation. As a consequence, there is no need to think that the theorem requires endorsing the atomism of voter deliberation that some commentators have attributed to Condorcet on the basis of a strict reading of the independence assumption (Grofman and Feld 1988). In fact, if one looks at Condorcet's more general philosophy, it is obvious that deliberation is not excluded from his ideal scheme. The right way to read him, therefore, is to consider the independence assumption an idealization of a low vote correlation made possible in a free and plural society where a diversity of views are encouraged, rather than interpret him as a Rousseauian democrat willing to ban pre-voting deliberation among voters.[9]

From that perspective, deliberation can in fact be a good thing if it increases the judgment accuracy of the average voter. The problem for the independence assumption is not communication among voters,

[9] This anti-deliberation stance cannot even be attributed to Rousseau without qualification. See Waldron 1989.

information sharing, or any kind of pre-voting deliberation (Estlund 1989: 1320). It is blind deference to someone else's authority. In fact, it is not even clear that the presence of mutual or common influences such as opinion leaders rules out independence. Indeed, under certain conditions, deference to opinion leaders can improve individual competence without violating independence and can thus raise group competence as well (Estlund 1994).

The assumption of independence in the CJT empirically translates as the flip side of the assumption of "enlightenment"—not the cognitive side tied to education but the agentive side tied to autonomy. On that reading, independence does not mean the absence of interaction with other people's opinion or a lack of exposure to their ideas but rather the ability to understand their ideas and critically reflect upon them. These abilities actually *require* exposure to diverse ideas, which is made possible by a free press, for example. But this ideal also demands a capacity of citizens themselves to make up their minds autonomously, thus pointing out the importance of a certain kind of education. While the prerequisite that such an education be widespread in society may seem utopian, it suggests that the independence assumption is no more demanding than the ideal of a liberal society.

1.4 Sincere Voting

The last assumption is that of sincere or truthful voting. Truthful voting should be defined in opposition to both self-interested and strategic voting. While self-interested and strategic voting are often confused, they raise different problems for the Condorcet Jury Theorem.

In the theoretical framework of the CJT, voting is a judgment about the common good, not the expression of a self-interested preference or even a judgment about what is most likely to best serve *me*. For a long time, economists and rational choice theorists in general have taken the assumption of self-serving voting for granted and the assumption of altruistic voting as utopian. The literature on voters' psychology, however, suggests that altruistic voting is more plausible than self-interested, pocketbook voting (e.g., Popkin 1994: 21). In fact, in light of overwhelming empirical evidence in favor of non-self-interested voting, rational choice theorists are now arguing that it is rational to vote non-selfishly, that is, altruistically or ideologically (e.g., Bowles and Gintis 2006; Caplan 2007).[10] If my vote makes little difference to the collective outcome, I can

---

[10] For the original formulations of the critique of rational choice theorists' defense of self-interested voting, see, e.g., Goodin and Roberts 1975 and Brennan and Lomaski 1993. See also Elster 2009 for a demonstration that the economic axiom of self-interest—applied to voting or other behaviors—is often psychologically implausible and, in fact, not even essential to economic theory.

indulge in ideological or expressive voting and feel good at hardly any cost to myself.

Truthful voting is also distinct from strategic voting. In truthful voting, voters do not seek to hide their genuine preferences and do not manipulate their votes either to serve their interest (as in pocketbook voting) or others' interests (as in altruistic voting) or to throw off the result (as in what could be called "antisocial" voting). The Condorcetian assumption of nonstrategic voting seems far-fetched to social choice theorists, given the incentive structure of voting systems. Truthful voting does not make sense from a rational choice point of view (e.g., Austen-Smith and Banks 1996; Fedderson and Pesendorfer 1997).[11] This is so, in particular, because even in a group sharing a notion of the common good, "sincere voting does not constitute a Nash equilibrium" (Austen-Smith and Banks 1996: 34). In fact, in the jury context, the probability of attaining the truth is actually maximized when voters vote strategically (e.g., Dekel and Piccione 2000). The pursuit of truth at the level of the group is not contradictory with strategic voting where there is a common goal.

In relation to that question of strategic versus sincere voting, we should observe that the problem of agenda manipulation does not arise for binary options, so the Condorcet Jury Theorem itself is safe from the possibility of strategic voting. It could be, however, that agenda manipulation occurs during the antecedent phase, when the options to be presented are decided on. While this does not affect the validity of the theorem and the epistemic properties that can be expected from majority rule, it certainly raises questions as to whether majority rule can ever be proved useful when it is so dependent on a preselection of options that is itself vulnerable to strategic manipulation. Let us imagine that there are three options: A, B, and C. Option A is objectively the best, option B is tolerable, and option C is the worst. What the CJT guarantees is that for whatever binary choice, the majority will prefer the better option over the worse one. But offered a choice between B and C, majority rule has no chance of selecting A. This is not a weakness of majority rule per se but is nevertheless an issue to be considered. As suggested above, one way to answer that problem of agenda manipulation is to let chance, rather than any possibly strategic individual decide on the agenda. This introduces a certain amount of indeterminacy in the final outcome—option A could still fail to be selected—but indeterminacy is better than the systematic epistemic loss guaranteed by strategic manipulation. Another solution

---

[11] For Austen-Smith and Banks, the assumption of sincere voting "is inconsistent with a game-theoretic view of collective behavior. A satisfactory rational choice foundation for the claim that majorities invariably 'do better' than individuals, therefore, has yet to be derived." (Austen-Smith and Banks 1996: 35).

would be to have the agenda set by an elected representative who is held accountable to the deliberating assembly.

More generally, truthful voting seems plausible if one remembers that strategic voting is problematic in a context where strategic interactions matter, such as a small-sized jury. However, the interesting application of the jury theorem for democracy is at the limit, where large numbers are involved and strategic interactions disappear. There, the assumption of truthful voting is at least theoretically supported. In mass elections, since one's vote does not make a difference anyway and assuming that the opportunity costs of voting are not too high, it becomes rational to vote truthfully (Austen-Smith and Banks 1996; Feddersen and Pesendorfer 1997). Thus, at the very least, the CJT's assumption of truthful voting has yet to be disproved.

Even if Condorcet's assumptions can be defended as plausible and intuitive to some extent, they also point out the dark side of majority rule. If the probability that the average voter is right on any yes-or-no question is lower than .5, then majority rule is virtually certain to produce the wrong outcome. If votes are not sufficiently independent or people do not vote sincerely, the same thing happens. When and where these assumptions are likely to be violated, therefore, the rule of the many seems like a terrible idea—worse, in fact, than any version of the rule of the few. On some issues, it might thus seem preferable to replace majority rule with "supermajorities" or nonelected institutions, such as the Supreme Court. Defining the domain where majority rule is likely to produce worse outcomes than the rule of the few is partly a matter of experience and knowledge of history and partly a gamble. I turn to the domain question in the last section of this chapter.

2. THE MIRACLE OF AGGREGATION

The "Miracle of Aggregation" (Converse 1990; Caplan 2007)[12] is another explanation for collective intelligence distinct from the Condorcet Jury

---

[12] Another, more popular origin for this name might be found in an article from the *New York Times* published on March 11, 2006, titled "The Future Divined by the Crowd." In this article, the journalist Joe Nocera marveled at the accuracy of prediction markets—those markets in which groups of people guess or bet on something with the results aggregated into a consensus. The journalist concluded his article by using the stock reply given by the character of the producer in the movie *Shakespeare in Love* when asked how it was that a play that seemed to be such a hopeless muddle during rehearsal was transformed into a gem on opening night: "It's a miracle." Nocera thought that this expression described prediction markets perfectly. As it turned out, the original line from the movie was "It's a mystery," not "It's a miracle." But somehow the label "miracle" used to describe the logic of the wisdom of crowds stuck with later commentators.

Theorem, although it is also dependent on the law of large numbers.[13] Although not a defense of majority rule per se, it is often invoked in defense of democracy, so let me try to connect the two.

The Miracle of Aggregation accounts for why the average guess of large groups of people on matters with a factual answer tends to be uncannily accurate, as in statistician Francis Galton's famous experiment regarding a weight-guessing competition. To rehearse briefly the now-worn anecdote, Galton was attending a country fair, in which one of the attractions was a guessing game.[14] The goal was to guess the weight of an ox once slaughtered and dressed. Galton took the answers of the 800 participants or so and computed the average, which turned out to fall within one pound of the right answer. Many other anecdotes, recounted in both Surowiecki (2004) and Sunstein (2006), vividly illustrate the same "miracle" in phenomena such as the predictive abilities of information markets on verifiable events such as Florida orange crops, movies' box office numbers, or election results. I will not consider information markets here, since they do not really qualify as democratic per se (see, however, appendix 3 for an exploration of their possible use for democratic decision making).

The most established version of the Miracle of Aggregation explains it as the statistical phenomenon by which a few informed people in a group are enough to guide the group to the right average answer, as long as uninformed people's answers are randomly or symmetrically distributed and thus cancel each other out.[15] Here collective intelligence actually depends on extracting the information held by an informed elite from the mass of "noise" represented by other people's opinions.[16]

As long as a large enough minority knows the right answer and all the others make mistakes that cancel each other out, the right answer is still, so to speak, going to rise to the surface.[17] The Miracle of Aggregation usually applies to cases where different individuals submit their own individual estimates of some continuous quantity (e.g., the weight of an ox in Galton's famous experiment) and the average is taken. In such situations, the options are many different possible values of the quantity in question;

[13]I take the Miracle of Aggregation to be the statistical version of what Condorcet's mathematical theorem formalizes, but I might be wrong in this assessment.

[14]The anecdote is reported by Galton himself in his *Memories of My Life* (Galton 1908: 246). For the scientific report in *Nature*, see Galton 1907a and 1907b.

[15]This "elitist" version probably goes back to Berelson, Lazarsfeld, and McPhee (1954). My thanks to Benjamin Page for helping me to figure out this distinction between the elitist and democratic versions of the Miracle of Aggregation.

[16]Notice that this explanation is less convincing when applied to the example of the guessing contest. This inadequacy does not seem to have struck Surowiecki or Sunstein.

[17]I say "large enough" because the minority in question needs to have high chances to be pivotal for the miracle to work. If the minority is too small, its judgment will be lost in the noise of wrong answers.

thus the options are continuous and potentially infinite and are not limited to two. The miracle can nonetheless be made to apply to binary situations too. To go back to the example developed in the introduction, let us say that the group lost in the maze is trying to decide whether to go left or right. The group contains a sizable minority who know the right answer (but are unable to convince the others through deliberation that they are right). All the other members of the group choose randomly, which results in a roughly equal split of those voters between the option "left" and the option "right." If the group resorts to majority rule, then the votes of the knowing group can be expected to tip the collective choice in favor of the right answer. This is a better option than flipping a coin, where there would be only a 50 percent chance of getting the right answer. It is also better than letting one randomly chosen member of the group make the decision, since the probability that a knowing person would be randomly picked would be $(1/n)^*m$, where n is the number of members in the group, m the number of persons in the knowing minority, and n is much greater than m.

A more democratic version of the Miracle of Aggregation presents things slightly differently. In this telling, everyone has an opinion that is roughly correct and the distribution of errors around each individual's "blurry" judgment is such that individual errors cancel each other out in the aggregate; thus the collective judgment is fairly accurate. In the example of the weight-guessing contest, this means that most people were not that far off the right weight, although none of them knew it exactly. Page and Shapiro (1992) apply this model to account for the rationality of public opinion. According to them, people have meaningful opinions surrounded by noise, and aggregation across individuals produces an aggregation of these real opinions. For example, some citizens underestimate and others overestimate the benefits of immigration. "Even if individuals' responses to opinion surveys are partly random, full of measurement error, and unstable, when aggregated into a collective response—for example the percentage of people who say they favor a particular policy—the collective response may be quite meaningful and stable" (Page and Shapiro 1992: 41). What Page and Shapiro imply is that the public is epistemically more knowledgeable as a whole than any of the individuals who make it up, which is why politicians are right to promote policies based on the public's judgment (a reasoning extended by Page and Bouton (2006) to foreign policy as well).

A third version of the Miracle of Aggregation sees the right answer dispersed in bits and pieces among many people. As long as people express a judgment that contains the accurate piece of information that they know and a random opinion about the piece of knowledge that they lack, the same logic of cancellation of random errors is still going to produce the right prediction in the aggregate. This explanation is unlikely to apply to the weight-guessing contest example—but if it did, it would require that some people in the group knew the weight of the cow's tail, some other people the weight of the ears, etc., and that they randomized their guess about the other parts. On average, all the pieces of information would aggregate to the right answer.

The Miracle of Aggregation, in its elitist, democratic, or distributed version, is an appealing way to account for the epistemic properties of democratic decision making. In effect, Galton himself, though not thinking very highly of democracy, was prompted by his own result to compare the gambling situation with democratic voting and to conclude that "[t]his result seems more creditable to the trustworthiness of a democratic judgment than might have been expected" (Galton 1907b: 451). For some theorists, the Miracle of Aggregation is an even better explanation for collective intelligence and why democracy works than the more traditional explanation in terms of deliberation and the pursuit of rational consensus.[18]

The Miracle of Aggregation can be made to apply to majority rule's epistemic properties, provided there are a few modifications. In the accounts surveyed above, the miracle occurs at the level of aggregated judgments about some quantifiable value—like the weight of an ox, the number of beans in a jar, the percentage by which a candidate will win or lose an election—for which people's guesses will be discrete and different values. For majority rule, where we assume that the answer is yes or no as opposed to such quantitative values, the aggregated judgment simply means whatever answer—yes or no, left or right—gets at least $(n/2)+1$ votes. If we assume that only one person in the crowd knows the right answer and the others randomize their choice, the miracle will occur only if the person with the right answer is also pivotal. As n grows large, the probability that the random guesses of the other $n-1$ persons cancel each other goes to 1. Conversely, however, the probability that the person with the right answer is pivotal goes to 0. As a result, the probability of the group getting the right answer if only one person in the group knows it remains very close to .5, neutralizing any hope for a "miracle" as numbers grow larger. So the Miracle of Aggregation cannot support the epistemic properties of majority rule in the case where only

---

[18] Cass Sunstein, for example, sees it as a "Hayekian challenge to Habermas" (Sunstein 2006). In fact, it is unclear both that the Miracle of Aggregation is the same thing as the invisible-hand mechanism at work in the emergence of the prices of goods or information in markets and that democratic deliberation is made superfluous by information aggregation through majority rule, polls, or markets.

one person in the group knows the right answer. It can, however, support more optimism if we assume that at least a sizable minority knows the right answer—sizable enough to have a better than random chance to be pivotal—or, as in the Condorcet Jury Theorem, that everyone in the group has at least a slightly better than random chance of knowing the right answer.

Assuming either a random or symmetrical distribution of errors around the right answer in the case of a sizable, knowledgeable minority or a more optimistic assumption of minimal enlightenment on the part of voters runs into the objection that it might very well be the case that, in fact, voters are systematically biased in the same direction, so that their collective judgment is going to amplify individual mistakes, not to correct for them. I postpone dealing with that objection until the end of this chapter. Let me now introduce a third alternative account of the epistemic properties of democratic judgment.

3. MODELS OF COGNITIVE DIVERSITY

Scott Page's model of the collective wisdom of group predictions (Page 2007; Hong and Page 2012; Lamberson and Page forthcoming) offers such an alternative account of the wisdom of crowds—a model whose conclusions can arguably be applied to the use of majority rule in a binary choice framework, as being a subset of the larger case of judgment aggregation over a continuum of options. Page's account is compatible with the probabilistic and statistical models (Condorcet Jury Theorem and Miracle of Aggregation), but it translates more satisfyingly into plausible empirical conditions. Most crucially, the emphasis in Page's account is not so much on the existence of a large number of votes as it is on the existence of sufficient *cognitive diversity* in the group, no matter its size. Cognitive diversity, as reviewed by earlier chapters, is roughly the fact that people make predictions based on different models of the way the world works or should be interpreted. Cognitive diversity ensures that votes (or predictions) are not independent but, on the contrary, negatively correlated. The good thing about negative correlation of this type is that it guarantees that where one voter makes a mistake, another is likely to get it right, and vice versa. In the aggregate, therefore, mistakes cancel each other not randomly but systematically.

The logic of cognitive diversity in group judgment aggregation relies on two main mathematical results: the Diversity Prediction Theorem and the Crowd Beats Average Law. I will not enter into the details of the demonstration behind either of them but take them as a starting point for my reflection on the epistemic properties of judgment aggregation through majority rule.

The first theorem states that when we average people's predictions, a group's collective error equals the average individual error minus their predictive diversity[19] (Page 2007: 208). In other words, when it comes to predicting outcomes (such as sales revenues or the weight of an ox), cognitive differences among voters matters *just as much as* individual ability. Increasing prediction diversity by one unit results in the same reduction in collective error as does increasing average ability by one unit.

Remember that in the case of deliberation applied to problem solving, we saw that cognitive diversity could actually trump individual ability. As long as there were enough people with different local optima, it did not matter if some people in the group had very suboptimal perspectives on the problem at hand. With aggregation, however, there is a strict tie between cognitive diversity and individual ability. This is because unlike what happens in deliberation, the better information, idea, or argument does not crowd out the worse. Judgment aggregation aggregates everything, including the bad input, which occasionally makes the group less smart.

The second theorem—the Crowd Beats Average Law—states that the accuracy of the group's prediction cannot be worse than the average accuracy of its members. In other words, the group necessarily predicts more accurately than its average member. Further, the amount by which the group outpredicts its average member increases as the group becomes more diverse (Page 2007: 197). This law follows directly from the Diversity Prediction Theorem.

Let us illustrate the model with the example of an election between two presidential candidates. The point of voting in such an election is, among other things, to identify who is the fittest candidate for office. Individually, each of the voters will make a prediction based on a limited number of factors: some of us will base our judgment on how competent with social issues a candidate is likely to be. Others will make a prediction based on both how fiscally conservative he is and what the state of the economy is presumed to be in the coming years. Still others will make a prediction based on a mix of factors: the candidate's charisma, the current price of oil, and the prospect that Iran obtains the nuclear bomb (I develop an idealized example in appendix 2 of this chapter). When looking at the candidates, we will thus look at different dimensions of the same quality (or in Page's vocabulary, "perspective"), in that case competence for office. This produces what Page calls "nonoverlapping

[19]Predictive diversity is, technically, the averaged square distance from the individual predictions to the collective prediction.

projection interpretations," that is, interpretations of the candidate's competence that do not contain any of the same variables or dimensions (e.g., competence on social issues or on economic issues).[20] The beauty of having such different predictive models in a group is that, because of the negative correlations between predictions they entail (see Hong and Page 2009 for a demonstration),[21] the group makes even better predictions than the CJT or the Miracle of Aggregation would predict (for binary predictions). In fact, as the group gets larger, the individual accuracy of predictive models matters less and less, while the cognitive diversity of the group of models matters more and more. In other words, "as groups become large, the criterion becomes less 'is this person accurate' than 'is this person different'" (Lamberson and Page forthcoming: 20).

This amazing result applies to the aggregation of predictions over a continuum of options, but there is no reason why it could not be extended, with the proper restrictions, to predictions over discrete binary options as well. Imagine that people are making predictions about the future of the economy that take the form of a continuum of values on a scale from 0 to 10. Their aggregated prediction is, provided the conditions of Page's theorem hold, more likely right than that of a random person in the group. Let's imagine this prediction is 4.5. Now imagine that any person whose predictive value is above 5 votes for candidate A and that any person whose predictive value is below 5 votes for candidate B (people predicting exactly 5 are randomly assigned to one candidate or the other). The majority's choice—in that case B—is a rough proxy for the collective prediction about the future. Conversely, the aggregated prediction about the future was an accurate predictor of which candidate would win. In other words, to the extent that the choice of a majority when choosing between two alternatives can be compared with the aggregated predictions of a group over a continuum of options (and even if the two cases don't overlap perfectly neatly, there is a vast number of cases in which they do), what Hong's, Lamberson's, and Page's results suggest is that we are better off with the median answer of a sufficiently cognitively diverse group of people than with the answer of a randomly chosen member of this same group.

[20] Page formalizes the "Projection Property" as follows: "If two people base their predictive models on different variables from the same perspective (formally, if they use nonoverlapping projection interpretations), then the correctness of their predictions is negatively correlated for binary projections" (Page 2007: 203).

[21] The gist of the paper consists in demonstrating that "seeing the world independently, looking at different attributes, not only does not imply, it is inconsistent with, both conditional independence of signals and independently correct signals" (Hong and Page 2009: 18). In other words, except in one very implausible scenario where all reasonable informed individuals each ignore a different piece of information, their predictions will not be independent but negatively correlated.

To recapitulate, the argument is that in order to maximize our chances of picking the better of two options, we are better off taking the median answer of a sufficiently cognitively diverse group of people than letting a randomly selected individual in that group make the choice for the group. This is because, for a given group of people using different predictive models, the predictions will be negatively correlated and mistakes will cancel each other not randomly but systematically. As a result, the average mistake of the group will be less than the average mistake of a randomly selected individual, and in fact, all the lesser as the difference between the predictive models used by those individuals is greater (i.e., as there is more cognitive diversity in the group). In the case of a vote between two options, this means that the median voter is more likely to be right than the randomly selected one, and all the more likely to be right as the group gets larger.

Hong and Page's account of the logic of group intelligence seems promising for an epistemic justification of majority rule, at least when majority rule is used in a group of people who make predictions based on different variables. A caveat needs to be added, however, lest the result seem too optimistic. One cannot have infinite variables or dimensions associated with a given perspective (say, competence for office). As the number of voters grows very large, the number of variables that people use to make a prediction may remain proportionally quite small (on top of social and economic issues, voters may look at personal charisma and foreign policy variables, but they might disregard variables such as dog type or sense of humor). To avoid positive correlations as the number of people in the crowd becomes larger, people must use cluster interpretations or they must base their interpretations on different perspectives. The interpretation used by someone who would judge a candidate both on his competence on social issues and his competence on fiscal issues is an example of cluster interpretation (see also the example of voter C in appendix 2). There is probably also an upper limit to the number of cluster interpretations that can be used, which would seem to mean that the cognitive model does not, at least not straightforwardly, support the epistemic properties of majority rule past a certain threshold, since the impossibility of producing new cluster interpretations may result in positive correlations between judgments that can harm the epistemic properties of judgment aggregation.

The superiority of a model based on predictive diversity over the CJT or the Miracle of Aggregation is at least twofold. First, the account gets rid of the awkward "independence assumption," which rendered both the CJT and the Miracle of Aggregation somewhat unrealistic (although not nearly as unrealistic as is commonly assumed) in their description of what is going on when people vote.

The second advantage is that this model supports the epistemic reliability of majority rule used among small groups.[22] Unlike what happens with the CJT or the Miracle of Aggregation, we do not need to have an infinity of voters for majority rule to guarantee 100 percent predictive accuracy. Because cognitive diversity can exist as soon as there is more than one person making the prediction, the magic can work for as small a group as three people (as in the admittedly contrived example above) and is substantially increased for any addition of a person with a sufficiently diverse predictive model to the group. In the CJT, by contrast, the major payoff of majority rule is only with very large numbers, and adding one person to the group does not make much of a difference. On the other hand, Page's account presents distinct advantages with large numbers, without failing to explain the epistemic payoff for cognitively diverse small groups as well. Thus, as the group gets larger, it is more important to have a cognitively diverse group than a group of accurate individuals.

This model, however, also suggests a diminishing return of adding people past a certain threshold, which will vary depending on the complexity of the prediction involved. This could seem slightly sobering from a democratic point of view favoring maximal inclusiveness, except if we assume, as I do here, that most modern democratic decision making takes place within assemblies that comprise fewer members than the point of diminishing returns. I think this is a fair assumption to make, given that the questions representatives usually deal with are complex and lend themselves to such a multitude of readings that every representative voting on an issue is more likely to contribute a prediction that is negatively correlated with those of his colleagues than one that is positively correlated. In other words, although Hong and Page's results do seem to speak against too much confidence in the epistemic properties of mass scale elections and referenda, the question of the possibly diminishing epistemic return of greater inclusiveness in voting need not arise for standard representative assemblies. In any case, I will assume here that this question of the diminishing epistemic return of greater inclusiveness in judgment aggregation is preempted by the issue of deliberative feasibility, which constrains *ex ante* the number of participants to the combined process of deliberating and then voting. The crucial point is that, as in the case of inclusive deliberation, the existence and cultivation of the right kind of cognitive diversity in the group is one condition for the collective intelligence of aggregated predictions to emerge.

This account of group competence, incidentally but importantly, also invalidates the main criticism that is often brought by deliberative democrats against judgment aggregation, namely, that a compilation of

[22] In fact, their account is more optimistic for small groups than very large ones.

discrete and partial views cannot add up to something meaningful. This criticism has been expressed by David Estlund, for example, in his treatment of the CJT (Estlund 2008: chap. 12). Estlund compares voters in the Jury Theorem scenario to blind men touching an elephant. The man who touches the leg thinks he is dealing with a tree, the man who is touching the trunk thinks he is dealing with a snake, and so on. No aggregation of such partial views, so Estlund argues, can produce the right picture of reality. In fact, what Hong and Page's model suggests, is that under some circumstances, the aggregation of partial and different views of the world can produce such a picture.

Of course, the model of group intelligence as a function of cognitive diversity is no more immune to the problem of systematic biases than the CJT and the Miracle of Aggregation are. If citizens share a number of wrong views—racist prejudices or the systematic biases diagnosed by Bryan Caplan (2007) in economic matters—majority rule is simply going to amplify these mistakes and make democratic decisions worse, if anything, than the decisions that could have been reached by a randomly chosen citizen. In Page's account, however, the risk of systematic mistakes can only happen if the group lacks both individual predictive accuracy—in other words, people's predictions are not sufficiently likely to be right—and diversity in the way individuals make predictions. Assuming minimally sophisticated voters relative to the questions at hand and a liberal society encouraging dissent and diverse thinking, however, one might argue that the worst-case scenario of a situation in which the average error is high and diversity low—the condition for the worst-case scenario of an abysmally unintelligent majority decision—is not very plausible.[23] Furthermore, Page's recent results seem to suggest that when the group grows large, individual sophistication or accuracy can be more than offset by group diversity. There is, of course, a lot of uncertainty as to when this kind of scenario might happen, but it supports a certain optimism even on issues where voters may not seem very reliable. I return to these issues in the next chapter.

What are the implications of Hong and Page's findings for majority rule? Majority rule—and, in fact, any democratic mechanism that aggregates individual judgments into collective judgments, such as polls—does have epistemic properties. Since the group's predictions are superior to those of the average citizen in the group, we have an argument for why the rule of the many is epistemically superior to the rule of one, when the one is randomly chosen. Notice, however, that this does not give us a maximal argument for majority rule, since majority rule by itself is not epistemically superior to the rule of a few smart people. Another of Page's

---

[23] For a more detailed critique of Caplan, see the following chapter.

results, the "Crowd's Possibly Free Lunch Theorem," might potentially provide that argument: roughly speaking, it establishes that a group of random people with diverse simple predictive models based on different interpretations of the world can occasionally predict a complicated function that beats in accuracy the function designed by expert statisticians (Page 2007: 234).[24] In Page's words, "simple, diverse predictive models can form sophisticated crowd-level predictions" and "these crowds can perform better than experts, provided their increased coverage more than makes up for the crudeness of their estimates" (Page 2007: 234). Such a "free lunch" is unlikely, however, so I will not rely on it.

In general, therefore, it is safe to assume that diversity does not trump ability when it comes to predictions. We might thus be better off as a group if a smaller group of smarter people made the decisions. Still, assuming democracy combines majority rule and inclusive deliberation, the Crowd Beats Average Law supports the argument that, overall, the rule of the many is epistemically superior to any version of the rule of the few.

APPENDIX 1. THE LAW OF LARGE NUMBERS IN THE CONDORCET JURY THEOREM

This series of graphs[25] illustrates how the probability that a simple majority is right goes to 1 as the number of voters grows large, provided the voters have a probability of being right on any binary question superior to 0.5. The formula used is $\sum n!/(n-k)!k! * (1-p)^{n-k} * p^k$.

Graph 1 illustrates how the law of large numbers operates for an average competence of .51, or 51 percent of being right on any binary issue.

Graph 2 provides a comparative view of how a slight increase in voters' competence quickly improves the competence of the majority. In this example, one can see competence increase from 50.5 percent to 52 percent by increments of 0.5 percent and see how the difference plays out as the group grows larger.

---

[24] An example might help. Let us say the goal is to figure out whether an incumbent to the presidency will be reelected. Obviously there are multiple factors to take into account to make that prediction. Most people, however, will rely on just one or two variables: some may use, say, the unemployment rate, while others may use the age of the incumbent and the level of oil prices. If we aggregate many people's predictive and sufficiently different models, they can occasionally add up to an extremely complicated function that could have been designed by a group of experts (except that the coefficients are not likely to be very accurate). If the group is lucky, this function yields more accurate predictions than those devised by experts themselves.

[25] Special thanks to Darko for designing them.

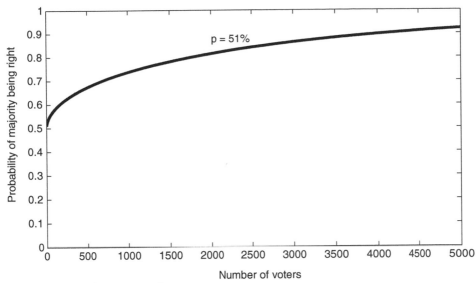

Graph 1. *Law of Large Numbers*

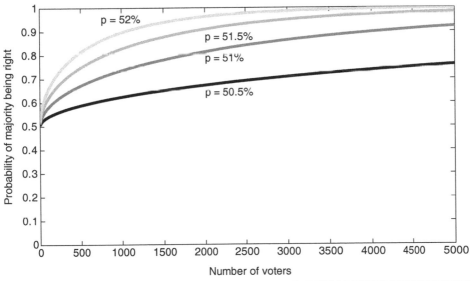

	10	100	1000	5000
0.5% (50.5%)	38.9%	50.0%	61.2%	75.6%
1% (51%)	40.2%	54.0%	72.6%	91.9%
1.5% (51.5%)	41.4%	57.9%	82.0%	98.2%
2% (52%)	42.7%	61.8%	89.1%	99.8%

Graph 2. *Comparative View*

Graphs 3–5 illustrate the role played by draws in situations where the group contains an even number of voters. The possibility of draws in such cases negatively and significantly influences the chances of the majority getting the right result, especially when the group is small (i.e., *n* is between 0 and 20). Graph 3 shows the probability of draws as the group

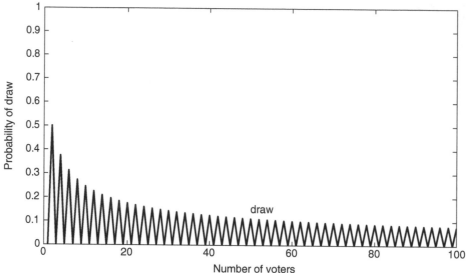

Graph 3. *Probability of Draws as Group Grows Larger*

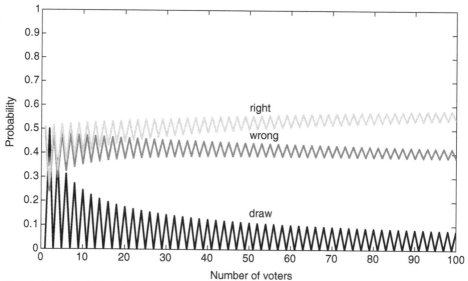

Graph 4. *Probability of Draw*

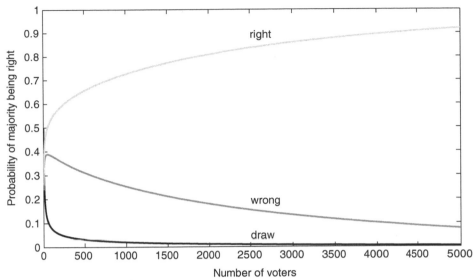

Graph 5. *Probability of Draw*

grows larger (the oscillation is due to the fact that draws are possible only for groups with an even number of members). These graphs illustrate—on the small and larger scale, respectively—what happens when the average probability of being right is 51 percent (upper curve) and when the average probability of being wrong is 49 percent (middle curve), taking into accounts the possibility of draws in each case (lower curve).

APPENDIX 2: THE LOGIC OF COGNITIVE DIVERSITY
IN JUDGMENT AGGREGATION

In this appendix, I develop an example illustrating how group predictive competence emerges from negatively correlated judgments.[26] Consider three voters A, B and C, who are trying to predict the competence or incompetence of several presidential candidates. We assume that they vote based on their predictions (sincere voting). As each person predicts either a competent (C) or an incompetent candidate (I), there cannot be any ties. Let us say that A is a Democrat, so he makes predictions based on how well he thinks the candidate will perform on social issues. Anyone at least moderately progressive is considered competent. B is a Republican, so he

[26] I built it on an example of Hong and Page (Page 2007: 199–205).

judges the candidate based on how fiscally responsible he thinks the candidate will be. Anyone at least moderately fiscally conservative is competent. C is an Independent, and he judges on a mix of both factors. He predicts that those candidates that are at least moderately fiscally conservative and either a little or moderately socially progressive will be competent.

Let us now assume that table 0 presents the mapping from candidates' attributes to whether the candidate would be competent (C) or incompetent (I). I offer this table as an arbitrary possible standard of the "right" answer in the choice of a candidate. Any other mapping could be imaginable.

Let us now consider table 1, which summarizes the predictions made by voter A. Note that 10 out of A's 16 answers turn out to be right (the answers in bold letters) when compared to the "reality" defined by table 0.

Table 2 summarizes the predictions of voter B, who turns out to be right 14 times out of 16 (in italic letters).

Table 3 summarizes the prediction by voter C, who is also right 14 times out of 16 (in underlined letters).

Table 4 summarizes the "agreement set" of A and B (in bold italic letters), which is designed so that where A and B agree, they are right.

Table 0: *The Mapping of the Candidate's Attributes to Competence as a President*

Fiscally conservative	Socially progressive			
	Highly	Moderately	A little	Not at all
High	C	C	C	I
Moderate	C	C	C	I
Low	I	I	I	I
Not at all	I	I	I	I

Table 1: *A's Predictive Model*

Fiscally conservative	Socially progressive			
	Highly	Moderately	A little	Not at all
High	C	C	I	I
Moderate	C	C	I	I
Low	C	C	I	I
Not at all	C	C	I	I

What happens when A and C disagree? There, C becomes the pivotal voter who determines the group's prediction. What happens then is striking. Looking at table 4, we see that A and B make different predictions for 8 of the boxes—the boxes not in their agreement set. Filling in C's predictions in those 8 boxes gives the predictions from the crowd shown in table 5.

Table 2: *B's Predictive Model*

Fiscally conservative	Socially progressive			
	Highly	*Moderately*	*Low*	*Not at all*
High	C	C	C	C
Moderate	C	C	C	C
Low	I	I	I	I
Not at all	I	I	I	I

Table 3: *C's Predictive Model*

Economically conservative	Socially progressive			
	Highly	*Moderately*	*Low*	*Not at all*
High	I	C	C	I
Moderate	I	C	C	I
Low	I	I	I	I
Not at all	I	I	I	I

Table 4: *A and B's Agreement Set*

Economically conservative	Socially progressive			
	Highly	*Moderately*	*Low*	*Not at all*
High	C	C		
Moderate	C	C		
Low			I	I
Not at all			I	I

Table 5: *The Group's Prediction, Using Majority Rule*

Economically conservative	Socially progressive			
	Highly	*Moderately*	*Low*	*Not at all*
High	C	C	C	I
Moderate	C	C	C	I
Low	I	I	I	I
Not at all	I	I	I	I

Table 5 summarizes the group's prediction—that is, the decision on which the majority of those three voters agrees. Note that table 5—the majority's prediction—is exactly like table 0—reality. This means that using majority rule, the group is able to predict accurately *every time*!

This example is, of course, carefully crafted to do the job, which is to illustrate how majority rule can produce more amazing results than even the CJT or the Miracle of Aggregation would predict.[27]

According to Hong and Page, the aggregation of predictive models does such a great job at producing correct decisions because of the existence of negative correlations between voters' predictions (in the CJT or the Miracle of Aggregation, by contrast, votes are supposed to be independent). I leave it to the reader to go back to the actual mathematical demonstration of the more general theorems (Diversity Prediction Theorem and Crowd Beats Average Law) in Page (2007: chap. 8). Let me, however, illustrate how negative correlations work in the example crafted above.

Take a minute to compare A and B's prediction tables (tables 1 and 2). B is right 14/16 (or 7/8) of the time. A predicts correctly in just 8 of those 14 times where B is right (i.e., 4/7 of the time). The result of 4/7 is less than A's actual score of 10/16 (or 5/8) of predicting correctly in general. If A's probability of predicting correctly were independent of B's probability of being correct, then A should predict correctly 5/8 of the time (which is 8.735 out of 14 cases). Since 8/14 is less than 8.735/14, this goes to show that A predicts correctly less often when B is right than would be the case if his predictions were independent of B's. In other words, A's and B's predictions are negatively correlated.

---

[27] As Page says, "the Law of Large Numbers [CJT] cannot get you to 100 percent and neither can canceling errors" (Page 2007: 202). Technically, you do get 100 percent accuracy with the CJT and the Miracle of Aggregation, but only at the limit case involving an infinite number of voters.

APPENDIX 3: INFORMATION MARKETS AND DEMOCRACY

> "No one in this world, so far as I know, has ever lost money
> by underestimating the intelligence of the great masses of the
> plain people."
>
> —H. L. Mencken

This appendix is meant to provide some background information and clarification about a market-based procedure for aggregating the wisdom of crowds, one that cannot offer a proper alternative to majority rule but is nonetheless very interesting in and of itself: information markets. Incidentally, information markets are also the best refutation I can think of to H. L. Mencken's smarty-pants quote above. The first section explains the nature and functioning of information markets; the second section explains how they relate to, and differ from, democratic decision procedures like voting; and the third section explains in what ways the market-based wisdom of the crowds can be used to supplement the democracy-based wisdom of the people.

1. Nature and Logic of Information Markets

Information markets, also called prediction markets, are markets in which groups of people guess or bet on something, with the results aggregated into a consensus. Whereas elections (or polls) are generally meant to aggregate people's preferences on some issue or for some candidates, information markets aggregate people's beliefs about some future outcome or event. Regarding election results, for example, information markets ask people what they think is the most likely outcome, whereas polls generally aggregate that answer from people's answers about how they say they will vote. In information markets, people may gamble their money on the election of candidate B and yet cast their vote for candidate A come the actual election. Of course, some polls also interrogate people about their beliefs about who is going to be elected as opposed to their voting intentions, but that is not the most common practice.

The oldest information market is the Iowa Electronic Markets, which have been run by the University of Iowa since the late 1980s and allow people to make election predictions. Since their beginnings, the consensus on that information market has almost always beaten the polling data. In the 2004 presidential election, for instance, the consensus not only steadfastly predicted a Bush victory (as did the polls) but came within 1.1 percentage points of the actual result. While not perfect, the predictions made by information markets are surprisingly accurate—and in fact, research has found, more accurate than individual experts or polls.

Doubtful to even some economists a few years ago, the efficiency of information markets has now been endorsed by no less than Kenneth Arrow (see Arrow et al. 2008).

Information markets predict not merely the results of presidential elections but almost anything with a factual answer: from weather in Florida, to prices of commodities, Oscar winners, movies' box office earnings, and more. Let me illustrate the logic of the information market in the case of election predictions. While there exist differently structured information markets, the basic logic remains the same.

Consider the 2004 presidential election. In that particular case, anyone could enter the market by putting some money into the common pool. For each dollar an investor put in, he or she received two contracts, one of which paid $1 if Bush won, and one that paid $1 if Kerry won. Once contracts were in circulation, participants could buy and sell them to each other at a trading site on the Web. If the going rate for Bush was 53 cents, for instance, then it meant that the market as a whole thought that Bush had a 53 percent chance of winning. Once the election results came out, participants cashed in their winning contracts from the pool—the more "Bush" contracts they had, the more money they made. Conversely, the more "Kerry" contracts they had, the more they lost.

This is the way the first markets run by the Iowa Electronic Markets worked. In addition to these winner-take-all markets, the Iowa project runs markets in which participants can bet on what share of the vote each candidate will receive. For example, you could also have made money if you correctly assessed that Bush was going to win by exactly 53 percent percent of the vote.[28]

Why does that system of buying and selling contracts about the possibility of an event work so well in determining the probability of that event? The theory is that the aggregated hunches of many people with money at stake are likely to be more accurate than the opinion of disinterested experts or whoever bothers to take a pollster's call.

There are, more specifically, four elements that help explain information markets' accuracy. These elements might be more or less overlapping, but I think it is important to distinguish them analytically. The first one is the Hayekian logic generally at work in any market—Adam Smith's "invisible hand." The second is the law of large numbers (the same law at work in the Condorcet Jury Theorem and the Miracle of Aggregation). The third is information markets' responsiveness to the intensity of people's beliefs. Finally, the fourth is the role of monetary incentives in ensuring that people display the private information they hold. Let me say a few words about each factor.

---

[28]This is not the real number. In reality, Bush won with 50.73 percent of the votes.

The Hayekian logic at work in information markets is the same that ensures that regular markets of goods or stocks "clear"—that is, reach an equilibrium at which the price of a good or a stock is exactly equal to, respectively, the marginal cost of production of the good or the profit-making capacity of the firm.[29]

Economists have theorized the functioning of this economic invisible hand as a decentralized calculator of prices. In fact, in an economist's account, the market of goods is a market of information, in which price serves as an indicator of everyone else's knowledge and intentions. The price of a good indicates information that the player (buyer or seller) does not know, and by deciding to buy, sell, or abstain at that price the player himself contributes his own partial knowledge to the market of information.

Most of that knowledge, interestingly, is tacit knowledge, which people are not fully aware of even as they are sharing it via price signals or using it when making a price decision. Prices, in themselves, are the results of that complex calculus that is the sum total of the tacit knowledge residing in bits and parcels within each individual. According to Hayek, price signals are the only possible way to let each economic decision maker communicate tacit or dispersed knowledge to each other, in order to solve the economic calculation problem.

If the market for goods is in fact a market where information gets traded through prices, information markets are markets where information is the traded and priced good. As to the mechanism that transforms a crowd of not-so-informed people into a quasi-omniscient group, here is how a Wall Street strategist explains the phenomenon: "All of us walk around with a little information and a substantial error term. And when we aggregate our results, the errors tend to cancel each other out and what is distilled is pure information."[30] This description fits exactly the profile of the distributed version of the Miracle of Aggregation studied in chapter 6, in which all of us know something about the world and randomize when it comes to the things we do not know about, so that in the

[29] Notice that while it is not always easy to determine whether the price of a good is really equal to its marginal cost of production, or whether that of a stock accurately reflects the economic fundamentals of the firm (both may be subject to irrational over- or under-evaluation), the advantage with information markets is that there is a short-term verification mechanism. At some point defined by the contract, the event has or hasn't taken place. Because of this short-term verification mechanism, there is less room for price bubbles.

[30] Michael J. Mauboussin, a well-known Wall Street strategist and an adjunct professor at Columbia Business School, asks every year his students to vote for the winners in 12 categories of the Academy Awards, prior to the results being known. Then, after the Oscars have been awarded, he tallies the results and compares the students' predictions with the winners (*New York Times*, March 17, 2006).

176 • Chapter Six

aggregate, when many people are involved, the errors cancel each other out and the right answer emerges.

The second key element in the success of an information market is, in consequence, the large numbers of people taking part in it. The accuracy of information markets depends indeed on how "thick" those markets are. The Iowa markets typically have hundreds or even thousands of traders. Economists generally expect these so-called thick markets to form better predictions than do thin markets, which have fewer traders. Their reasoning is that the more traders there are, the more information is potentially available and the more opportunities there are for trading.

Under the right circumstances, however, even thin markets can make accurate predictions. In Plott and Chen's experiments at Hewlett-Packard Laboratories, markets consisting of about a dozen employees predicted future sales better than the company's usual methods of market analysis (Plott and Chen 2002). Plott and Chen made up for the small number of participants by the care with which they selected them. To maximize the different sources of information available to the market, they chose people across a wide range of the company's departments. They also included some uninformed speculators, both to provide liquidity to the market and to provide watchful eyes against illogical market behavior.

The law of large numbers does not mean that everyone in the market has to be actively engaged in trading contracts. In fact, interestingly enough, the experiments run on the Iowa Electronic Markets show that only 15 percent in the crowd are correctly informed and contribute to driving the market to the right price. The other 85 percent are just "white noise" (Klarreich 2003). (This situation fits the nondistributed version of the miracle of aggregation discussed earlier in the chapter.) The important fact is that the market would not work without these 85 percent who, overall, probably lose money. Large numbers, or "thick" markets, are needed for the wisdom of the informed 15 percent of traders to emerge. That is why this wisdom is, ultimately, that of all the participants in the market, not just that of the few who have the correct information.

Information markets also work because they are responsive to the intensity of individual beliefs. Unlike what happens in polls or in voting—where each individual is given only one voice—here the more convinced you are, the more you should be willing to bet. By allowing for a differentiated weight given to more or less firmly grounded beliefs, information markets generally end up aggregating the most reliable information.

Finally, information markets involve monetary incentives, which ensure that people disclose the private information they hold. It seems that these incentives could be nonmonetary without harming the accuracy of prediction markets. Whether people are playing for money, virtual money, or reputation, what matters is that they have a good reason to divulge the information they hold. One could thus imagine similarly improving the epistemic properties of voting, polling, or even deliberation by introducing nonmonetary incentives in these practices.

Now that we have clarified the reason why information markets work, the more interesting question can be asked: how do they compare with political means of aggregating information?

2. Information Markets versus Democratic Procedures

One aspect of the superiority of information markets over certain democratic procedures (like deliberation or voting) is that they do better in terms of predictive performance. This is because the monetary incentives at play in information markets solve the "cheap talk" problem that often plagues majoritarian answers (as expressed in polls) and public debates. In information markets, people put their money where their mouth is. In comparison with deliberation, in particular, it seems that information markets are much less likely to be subject to hidden profiles or information cascades, that is, situations where the right answer fails to be discovered because people do not reveal all the information they have or situations where people converge on a conclusion and ignore evidence contrary to it solely because everyone keeps following the report of whoever spoke first.

The superiority of information markets in avoiding hidden profiles and cascades might also be due to the relative anonymity of the system. By contrast with deliberative settings, where the personality of the participants comes into play, information markets do not involve any actual encounters between people. The safety of electronically mediated communication probably adds to the incentive to disclose one's information. Not only is there something to be gained from it, but there is hardly any "social" cost associated with it, even if one's beliefs are somewhat politically incorrect or would be frowned upon as ludicrous or crazy by participants in a deliberative setting.

Information markets also present an advantage compared with a nondemocratic way of aggregating information, such as relying on experts. While experts may be just as competent as the markets, and perhaps occasionally more so, they can be biased by ideological considerations or political expediency. In theory and in practice, at least observed so far, information markets are not subject to these biases.

In particular, information markets seem remarkably immune to manipulation. At least in the case of the Iowa Electronic Markets, most attempts at manipulation were short lived and failed miserably.[31] For example, in

---

[31] "Miserably" being the operative word here, since the manipulators' attempts were not only unsuccessful but extremely costly for them.

the 2000 presidential market, several people opened accounts on the same day, and each invested $500—the maximum allowed—in Pat Buchanan shares. Buchanan prices briefly spiked, but well-informed traders then seized the opportunity to profit off the manipulative traders and by the end of the day, the effect of the investments had virtually vanished.[32]

Second, information markets offer an automatic and impersonal way to distinguish between incompetent and competent bettors. They form a neutral and real-time test for better and worse information—since good information pays off and bad does not. In deliberation, to the contrary, rhetoric can fool other people, and people are not penalized for saying stupid things (at least rarely while the deliberation is going on). Indeed, in a deliberative setting, no one individual can legitimately claim to have the authority to silence anyone else in the name of his or her greater knowledge of the question at hand. In information markets, however, fools will rapidly lose a lot of money, without being able to blame the penalizing logic of the market as a personal attack on them.

Third, information aggregation in markets never ceases until the day when the event takes or fails to take place. Unlike people taking part in a deliberative session, the market never stops working, ensuring perpetual updating of the beliefs it reflects.

Information markets, however, are not perfect decision-making tools. For one thing, they cannot replace proper collective decision making on the ends to be pursued, which reveals a major limit of where they can be used in a democracy. Information markets cannot decide for us the degree to which we want opportunities to be equalized for all or the degree to which entrepreneurial risks should be rewarded, any more than they can help us decide on fundamental issues such as a right to abortion, capital punishment, and preemptive war.[33] They can essentially help a deliberative assembly or voters determine the best means to those ends, by providing a "reality check" on the factual conditions for given normative policy goals (Sunstein 2006: 93).

Another intrinsic limit of information markets is that they seem to work only when applied to questions over which information is dispersed. They notoriously failed to predict whether Iraq had weapons of mass destruction, or who was going to be the next nominee on the Supreme Court (ibid.). Cascades are more likely to happen when the

---

[32] In an experiment in the same market, economists Koleman Strumpf of the University of North Carolina at Chapel Hill and Timothy Groseclose of Stanford University made random purchases. "The market would typically undo what we had done in a few hours," Strumpf says. "People weren't being fooled by our crazy investments." (All examples borrowed from Klarreich 2003.)

[33] Although see Abramowicz 2007 for a suggestion that information markets could theoretically be extended to normative questions as well.

information is concentrated in a few people as opposed to relatively dispersed among a large number of individuals. Failing to identify whether information is adequately distributed to render information markets useful, however, is a human limitation, not a flaw of information markets themselves. Notice also that the time limit built in information markets prevents huge mistakes, such as the emergence of bubbles characteristic of markets without a limited time horizon (like the real-estate market).

Finally, information markets raise moral questions regarding the nature of the issues people can bet about. Is it immoral to bet on the possibility of a terrorist attack, that is, to try to make money on the death of other human beings? Another has to do with the mixing of politics and money. Is it moral that people should be compensated for civically valuable information they could have made available for free?[34]

The former question is not purely theoretical and was answered in the negative in 2003, when the project of the Policy Analysis Market was abruptly brought to an end by critics who labeled it as morally abject. This project, sponsored by the Department of Defense, would have created a market in which participants could wager on events in the Middle East, such as the gross product of Syria in coming years or the political instability of Iran. When it was suggested that terrorist attacks and political assassinations could be added to the predictive possibilities, congressmen trashed the project. A more sensible middle ground, obviously, would have been to keep the principle of the Policy Analysis Market and show some sensitivity in its use and application.

There are finally a number of issues raised by information markets that touch on their possibly antidemocratic nature.

The first objection points out that it would be odd to support information markets as an alternative to deliberation when markets are not a political way to make collective decisions. Markets are indeed a way to dispense with political decision making altogether: they deprive people of the possibility of collective action. Where markets replace governments, the ensuing state of the world—a certain distribution of goods—emerging from the competition between private actors can, as a result, not be blamed on or credited to any individual in particular. The outcome defined by the markets can only be attributed to Smith's "invisible hand"—an unaccountable entity. Similarly, if information markets make a mistaken prediction, no one in particular could be held accountable for that mistake.

The point, however, is not to replace government by unaccountable, impersonal information markets but simply to let the people or their

---

[34]Notice that this is a distinct question from the empirical question of whether introducing monetary incentives crowds out civic spirit.

representatives use those information markets where they are likely to yield superior predictions to those of a majority (of citizens or representatives) and inform public debates with useful data. The government would still be held responsible for the policy measures it enacts, even if they are based on information-market predictions. Robin Hanson has been advocating for almost ten years a new system of governance that he calls "futarchy," in which voters would vote to say what they want and speculators on information markets, rather than traditional politicians, would say how to get it. The idea, according to Hanson (2007), would thus be to "vote on values"—that is, use popular referenda and parliamentary debates followed by a vote to settle the question of what kind of ends to pursue—and "bet on beliefs"—that is, turn over the choice of policies to information markets. The verification mechanism for the predictions of information markets would be the after-the-fact measurement of national GDP by elected representatives. Endowed with such a function, information markets are not an alternative to democracy altogether, but simply a tool at the service of democratic decision making. Contrary to what Hanson himself suggests, speculators on information markets would not be making the decisions for the people. Their predictions as to what is the best policy would simply be used within a democratic framework to improve the political decision-making process and the efficiency of public policies. "Futarchy" is not in that sense an alternative to democracy (or to oligarchy or dictatorship, for that matter) but an aid to decision making that is technically available to any form of government.

Another objection may be raised about the potentially corrupting effect on civic virtue of citizens' participation in information markets, where they find themselves financially rewarded for information that they could make available for free. In answer to this, one might reply that a system of monetary incentives rewarding people for disclosing privately held information is distinct from the morally dubious practices of bribery or buying votes. That people could make money out of their privately held information should be seen more neutrally as legitimate compensation for something that might indeed be costly to divulge. After all, even proponents of as civic an enterprise as Deliberation Day, Bruce Ackerman and James Fishkin, suggest that participants be compensated at least $150 a day for taking part in those yearly national deliberative activities (Ackerman and Fishkin 2004). True, this amount is meant to compensate for the trouble of participating, not for the information contributed in the collective discussions. But that might actually be a flaw in the project, as all teachers who grade students' class participation based on sheer presence in the classroom (rather than the actual quality of the contributions made) would know.

A third objection worries that information markets are, in fact, dominated by an informed elite. As mentioned earlier, it has been shown that in practice, only 15 percent of the participants on these markets truly matter to setting the price of a contract, while the other 85 percent are simply white noise. This would be a valid objection if there were a way to identify who the 15 percent of informed traders are that did not itself rely on the market. But as far as we know, there is no way to do this. Furthermore, because of the distributed nature of the information at stake, the 15 percent are unlikely to be made up of the same exact people for every single question. Thus the group identified after the fact as competent on the question of a presidential election may have nothing to say about the probability of a bad orange crop in Florida or the price of gasoline next month. Therefore, even a post hoc sociological analysis of who the informed voters were on a given question would tell you nothing of value for the next question. There is no way around involving the many in information markets, even if for each question only a small fraction among them really knows what they are doing.

The objector might then wonder why 85 percent of the people are willing to be the honey that draws in intelligent traders, allowing that these smarter bettors are going to win at their expenses. One answer is that 100 percent of the bettors probably think that they know enough to profit from that knowledge. Overconfidence is rendered possible by the fact that in information markets, as in politics more generally, there is no way to tell in advance who knows and who doesn't. The only safe prediction is to bet that the group as a whole knows best. On an individual level, this means that most participants will continue to believe, irrationally for many, that they have a good reason to participate. Information markets work in spite of—or, rather, thanks to—many participants' overconfidence in their predictive power.

Information markets' responsiveness to belief intensity also introduces a seemingly undemocratic element. Information markets allow for the confidence in one's beliefs—their "intensity"—to matter and to weigh more heavily in the aggregated outcome. In the market, the more informed you are, the more capable you are of betting, and therefore the more willing you will be to bet. How are information markets democratic, then, if they violate the principle of "one man, one vote" or, more exactly here, "one man, one bet"?

The difficulty comes here from the fact that votes cast in an election are not the same as bets placed on a market. Votes simultaneously express judgments about values, immediate preferences, and beliefs about facts. It is often very hard to distinguish among the three components in a given vote. Notice that this might explain why a democrat like John Stuart Mill, though a democrat, did not endorse the principle of "one man, one vote" and proposed instead a rather complicated scheme of plural voting, which gave more weight to the knowledgeable votes. In his view, voting was about aggregating judgments and factual beliefs, not just preferences. So, while universal franchise minimally ensured that everybody's preference was equally taken into account, a greater number of votes for people with a higher ability to gather facts and make value judgments was also in order. But of course, preferences, factual beliefs, and value judgments are all lumped together in the right to vote. Though it often seems that Mill gave more weight to the preferences and interests of some citizens than to those of others, all he did was to ensure that a greater weight was given to the beliefs and value judgments of people with greater knowledge. In that way, Mill's scheme of plural voting was not antidemocratic. It was merely an attempt at increasing the epistemic quality of the democratic outcomes. However, in Mill's idea, the number of votes attributed to different people was indexed on the assumed superiority of their judgment, itself measured by the number of degrees they possessed or their results on standardized tests, not on the sheer number of dollars they were ready to gamble, which is all that information markets are sensitive to.

Bets should theoretically express only beliefs about certain facts (because if your value judgments or preferences start influencing your decisions on the market, you will probably lose money). While it is plausible that democracy requires that everybody's strict preferences be given an equal weight, no such thing obtains for individuals' factual beliefs, and perhaps not even of value judgments. If one thinks of deliberation, as opposed to voting, equality there requires equality of access and voice, not necessarily equality of influence on the outcome. The force of the better argument and the better information should triumph, which means giving greater weight to those with the most persuasive arguments and the better information. From a comparative point of view, information markets are no worse than deliberation in giving better information greater weight.

A last objection bears on the problem of potential manipulation by the rich, who can exert greater influence on the outcome. This objection encompasses two distinct risks associated with the purely monetary sensitivity of information markets. First, there is the problem posed by the irrational beliefs of rich people who can afford to gamble away all their money on the wrong outcome. Second, there is the problem of strategic manipulation by the rich, who may want to bias the predictive outcome of the information market (perhaps because they know that the result is taken into account to make political decisions later on).

Some have suggested a cap of five hundred dollars to avoid manipulation by the rich.[35] The risks attached to both the irrational beliefs and the willful manipulation of a few rich traders, however, seem empirically

---

[35] A surprising fact is that existing information markets apparently do not have real money at stake. As already said, stakes matter but they need not be monetary.

negligible. As the Iowa Electronic Markets repeatedly demonstrated, irrational fads have been temporary and willful attempts at manipulating the markets systematically have failed. There will always be more people willing to take advantage of the "mistake" induced by wishful thinking or outright manipulation than delusional or manipulative traders themselves. Manipulation of the market, in particular, is not only incredibly costly but ultimately inefficient.

3. Possible Role for Information Markets in a Democracy

What are the institutional implications suggested by these analyses? In other words, how can we harness collective intelligence, which information markets seem to channel, to the benefit of smarter democratic decision making?

The first practical implication, or so it seems, would be to revive the defunct Policy Analysis Market and, more generally, to use such information markets to inform policy. This would not be the first time, in fact, that democracies resort to information markets. According to Josiah Ober (2010), the ancient practice of ostracism, by which Athenian citizens were asked to identify through a vote the prominent political figure most likely to become a threat for the city, exiling the "winner" for ten years as a consequence, was a sort of proto–information market. Justin Wolfers and Eric Zitzewitz (2004) give an example of how useful information markets could be with the case of what they christen a "Saddam security," measuring geopolitical risk through predictions on oil prices. The gist of these authors' idea is to allow the aggregated expertise of the people to inform policy decisions in real time. They argue that information markets represent a useful supplement to decisions by experts who tend to be biased by ideological predisposition and to focus on the most analyzable costs of a policy. In Robin Hanson's "futarchy" scheme, the use of majority rule would be restricted to questions of value and fundamental preferences, whereas information markets would be used to deal with beliefs, or what I call more generally predictions (Hanson 2007). This is also an institutional reform advocated by Sunstein (2006).

These suggestions are interesting, especially in light of our previous discussion about the ambiguity of the content of a vote. If votes could indeed be restricted to expressing preferences and pure value judgments, as opposed to also expressing factual beliefs, we would perhaps achieve the goal of a smarter democracy. Whether such a restriction is actually empirically feasible is another question altogether.

A second major institutional reform that the success of information markets suggests is the injection of incentives, financial or otherwise, to disclose private information into deliberative settings. In all empirical studies of deliberation, the major failings are "polarization" and "hidden profiles." If dissenters were more often rewarded and encouraged to bring to light unpopular information or arguments, the ultimate quality of the decision would no doubt be improved. The deliberative context should be such that dissenters are welcomed, respectfully listened to, and possibly even rewarded after the fact, if only symbolically, for disclosing information or arguments that later turned out to be particularly crucial.

CHAPTER SEVEN

Epistemic Failures of Majority Rule: Real and Imagined

IN THIS CHAPTER, I address a series of objections to the claimed epistemic properties of majority rule and, more generally, aggregation of judgments. The first section thus considers a general objection to the epistemic approach to voting, which supposedly does not take seriously enough the possibility that politics is about aggregation of interests, rather than aggregation of judgments. In this section, I also consider the objection from Arrow's Impossibility Theorem and the doctrinal paradox (or discursive dilemma). The second section addresses the problem of informational free riding supposedly afflicting citizens in mass democracies, as well as the problem of the voting paradox (as a by-product). The third section, finally, turns to a refutation of the objection that citizens suffer from systematic biases that are amplified at the collective level.

1. POLITICS OF JUDGMENT VERSUS POLITICS OF INTEREST AND THE IRRELEVANCE OF ARROW'S IMPOSSIBILITY THEOREM

In the previous chapter, I focused exclusively on an epistemic account of voting as the expression of a judgment, to the exclusion of the rival aggregative account based on the conception of voting as the expression of a preference or an interest. Let me now try to justify this choice and, more specifically, the rejection of the aggregative model inspired by economic theories of democracy. I see three problems with it.

First, if politics were about aggregation of preferences alone—and here I take preferences to mean, in the largest sense, any kind of preferences, whether selfish or altruistic—the only way we could come up with an argument for dictatorship or oligarchy rather than for democracy is if we thought it important to satisfy the preferences of some individuals but not of others. The preference-aggregation paradigm, together with some minimal and widely accepted egalitarian assumptions that each individual's preference satisfaction is equally important, would immediately settle the case for democracy. The preference-aggregation model makes the justification of democracy almost too easy, since even an idealized dictator would have to resort to some kind of democratic procedure—a vote or a poll—in order to elicit the required information from all citizens about their individual preferences (see also Nelson 1980 on this point). Only if politics is about something other—and *more*—than pure preference aggregation and equal preference satisfaction is the competition between rule of one, rule of the few, and rule of the many relevant and interesting.

Another problem is that the aggregative approach tends to interpret citizens' behavior along the same line as consumers' behavior. Whereas the concept of preference is theoretically content-neutral, the aggregative democracy literature often fills it with the more restrictive notion of "self-interested preference."[1] In that view, voters consume policies the same way that consumers buy goods at the supermarket, maximizing a self-interested utility function. This approach to politics, however, has only a partial empirical validity, since on many questions people do not vote their self-interest (see even Caplan 2007 on that point).[2] From that point of view, the contrast is between a politics of judgment, which requires an orientation toward the common good, and a politics of interest, where the interest at stake is, however informed and thought through, self-centered and individualistic.

Finally, it is possible that viewing collective decisions as a matter of interest or preference aggregation—rather than a matter of judgment, knowledge, and deliberation—makes them more vulnerable to the irrationalities studied by social choice theory and expressed by Arrow's famous Impossibility Theorem (Arrow 1953). According to this theorem, when more than two options are at stake, one cannot guarantee that there exists a social function capable of aggregating individual preferences (of a certain type) in a way that does not violate a number of axioms defining the conditions for the "rationality" of a collective choice.

As reviewed in chapter 2, Arrow's theorem has triggered a lot of skepticism toward the very possibility and meaningfulness of democracy as a procedure for aggregating preferences. This skepticism has been challenged empirically by critics who show that problems of cycling (as well as agenda control, strategic voting, and dimensional manipulation) are not sufficiently harmful, frequent, or irremediable to be of normative concern (e.g., Mackie 2003).[3] It has also been shown that Condorcet

---

[1] A point I develop at some length elsewhere (see Landemore 2004).

[2] One of the premises for Caplan's larger claim that voters are "rationally irrational" is his acknowledgment that the assumption of the selfish voter is wrong.

[3] Mackie, for example, dismisses as erroneous the historical examples adduced by Riker to illustrate the problems of cycling and instability (in particular, Riker's argument that the US Civil War was due to arbitrary dimensional manipulation).

cycles theoretically are not very likely in large societies (e.g., Tangian 2000; List 1998, 2001; Gehrlein 2002).[4]

Some deliberative democrats have argued that Arrow's theorem is not as damaging in the context of deliberative democracy, where a deliberative phase precedes the aggregation of preferences. Their strategy consists in questioning the relevance of Arrow's assumptions for the domain of deliberative politics (where politics is deemed to be about judgments about the common good), by contrast with aggregative democracy.

The main target, in that respect, is the assumption of "unrestricted domain," which has been read by critics as requiring that the social choice function account for all preferences among all voters (including, for example, selfish, racist, or antisocial preferences if they exist). Critics have suggested that this assumption is too demanding, since in any society preferences are de facto restricted (Mackie 2003), and even if they were not, one can make a reasonable case that they should be. Joshua Cohen (2004), for example, argues that the input that should enter the social decision function should be constrained and passed through a Rousseauian or Rawlsian filter to ensure that only "generalized" rather than selfish or racist interests, and "public reason" rather than ideological or self-serving types of arguments, are ultimately aggregated. Cohen thus argues that there is no normative reason not to accept a relaxation of the axiom of unrestricted domain—a relaxation many liberal democracies actually live with (e.g., banning racist preferences). If we aggregate preferences over reasonable social policies rather than ice cream flavors, in other words, the problems diagnosed by Arrow shouldn't arise.

This kind of objection, however, fails to mention or recognize that the unrestricted-domain assumption is not an expansion condition. It does not say that any desire of any citizen (qua any policy she wants included) must be in the feasible set. It simply says that once a feasible set of alternatives (expressed as preferences or judgments or desires or whatnot) is specified, then any pairwise ordering of that set must be admissible. So the problem diagnosed by Arrow might well arise even if we filtered down preferences to only three publicly minded policy choices, just as they could arise for three ice-cream flavors. Another way to say this is that plugging judgments about the common good rather than narrow interests and preferences in the aggregation function does not change

[4] List shows that contrary to what standard results suggest, the probability of cycles need not increase as the number of options or individuals increases as long as systematic deviations, however slight, from an impartial culture situation—in which any imaginable possible preference is represented—can be assumed. Since no actual democracy has an impartial culture, the probability of cycles in such a society will typically converge to zero as the number of individuals increases.

anything, as judgment aggregation is merely a special case of Arrowian preference aggregation.[5] Reducing the domain of judgments to a narrower set of normatively acceptable views would not necessarily solve the fundamental problem pointed out by Arrow.

The fact is that Arrow's theorem can be generalized to the case of a "restricted domain" of preferences as long as, for that restricted domain, certain kinds of preferences are possible inputs. The belief that the set of immoral, wrong, or selfish preferences weeded out by the deliberative process will exactly overlap with the set of preferences that are problematic from a social choice point of view is not fully theoretically supported. In other words, there is no absolute guarantee that among the judgments that are normatively acceptable, there do not exist judgments that give rise to majority preference cycles.

There is nevertheless an intuition behind deliberative democrats' initial objection that might well be correct: the idea that a successful deliberative process would help reduce the likelihood of Condorcet cycles—that is, would shape the set of preferences to be aggregated in such as way as to avoid the social theoretical problems predicted by Arrow. If that were true, then, indeed, Arrow's theorem wouldn't be as much of a problem for the aggregation of post-deliberative preferences as it can be for the aggregation of pre-deliberative preferences. Attempts at proving this promising conjecture have been made. For example, a group of political scientists running a field experiment within one of James Fishkin's deliberative polls recently showed that deliberation can reduce the risk of cycling majorities by making preferences more single-peaked (Farrar et al. 2010; see also List et al. 2006).[6] Unfortunately, a limit of these experiments seems to be that the pre-deliberative set of options already contain a Condorcet winner, so all the results prove is that once a Condorcet winner exists, deliberation brings preferences closer to it. They do not prove that deliberation can transform cyclical preferences into noncyclical ones, that is, turn non-single-peaked preferences into single-peaked ones. A lot of theoretical and empirical work thus remains to be done to support the promising conjecture.[7]

---

[5]For a technical dispute of that point, see, e.g., Dietrich and List (2007), who explicitly argue that judgment aggregation is not a special case of Arrowian preference aggregation but nonetheless prove an impossibility theorem comparable to Arrow's (and from which Arrow's theorem can, in fact, be derived) for what they argue is the distinct question of judgment aggregation.

[6]In this paper, the authors show that deliberation tends to bring the rankings of policies (the post-deliberative preferences) closer to single-peakedness, meaning that such preferences can be aggregated without the risk of producing cycling majorities.

[7]I thank Sean Ingham (Harvard University) and an anonymous reviewer for Princeton University Press for helping me clarify the meaning of the unrestricted-domain assumption.

To the extent that majority rule is never "merely majority rule," as Dewey wrote (Dewey [1927] 1954: 207), but is always inscribed in the larger context of a public debate, there are reasons to think that if deliberation has the property of reducing the likelihood of Condorcet cycles, impossibility theorems would not be as much of a practical threat to the value of majoritarian decisions as critics may have once thought.

The Arrowian objection is powerful and deserves to be taken seriously by deliberative and epistemic democrats. Yet I do not think it is enough to challenge the epistemic approach to democracy in general and the specific epistemic claim in favor of democracy made in this book. Whatever theoretical and empirical limits to majority rule this objection may rightly emphasize, those problems seem to occur at the margin, for the very specific case of cycling preferences. Until it is proven that the core of democratic judgments is generally "empty"—that is, that even post-deliberative democratic preferences are systematically, rather than rarely, cyclical—I think it is possible to proceed as if, for the most part, it made sense to try and aggregate individual judgments into a collective one.

Admittedly, even bracketing the issue of cyclical preferences and considering that, by and large, an epistemic approach is unaffected by the problem of cycling judgments, one runs into other problems, such as the doctrinal paradox, or discursive dilemma. The discursive dilemma characterizes the disagreement sometimes observed between the results of premise-based and conclusion-based decision procedures, and has been the object of an extensive literature in its own right (e.g., Kornhauser and Sager 1986, 1993; Kornhauser 1992; Chapman 1998; Brennan 2001; List and Pettit 2002, 2005a, b; List 2006). The disagreement between collective results achieved one way or the other arises for situations where we dichotomize between the premises and the conclusions of a given judgment.

The situations modeled by the discursive dilemma are not particularly realistic takes on what happens in actual elections, which generally proceed on a conclusion basis. More importantly, from a theoretical point of view, one can imagine that some deliberation about the best way to proceed—aggregation based on premises or aggregation based on conclusions—would take care of the indeterminacy issue, in which case the conflict need not arise. We commit in advance to one procedure or another and there is no point in comparing the results ex post. What would be more interesting is if it could be shown that one way of proceeding is more epistemically reliable than the other. As far as I know, however, the literature on the discursive dilemma is mostly concerned with formal problems of "coherence" between the aggregation of premises and the aggregation of conclusions, not so much with epistemic issues of "correspondence" with the truth or some procedure-independent standard of objectivity (see List 2012 for an illustration, and Goldman 2004 for a critique, of the coherentist branch of social epistemology). From the point of view of the epistemic case presented here, all that matters is that, whether the aggregation proceeds on the basis of premises or on the basis of conclusions, it is sufficiently epistemically reliable. While the literature on the discursive dilemma raises an interesting theoretical challenge, the impossibility theorems that it has engendered so far do not seem to me as threatening to the possibility of judgment aggregations as Arrow's theorem is for the aggregation of preferences about three options or more.

Most importantly for this book, in any case, it is possible to frame judgment aggregation in terms of binary options only, for which the Arrow Impossibility Theorem does not apply. To rehearse briefly the reasons adduced in support of that solution in the previous chapter: theoretically, one can imagine that pre-voting deliberation generates a partitioning of political problems into binary options. Should deliberation fail to reduce the choice to two options, a random mechanism of agenda setting could, for example, be used to determine the order in which pairwise voting between the multiple competing options will be organized.[8] Empirically, it is a fact that many assemblies break down their decisions into pairwise votes, with one option supporting the status quo and the other amending it.

Let me now go back briefly to the more substantive (as opposed to formal) aggregative approach to democracy, which emphasizes the fact that democracy is about aggregating preferences that are conceptualized as individual interests rather than as generalized interests or judgments about the common good. I said earlier that this approach, which reduces voting behavior to market behavior, is not empirically plausible. I would also argue that it is concerned with a question entirely orthogonal to that addressed by an epistemic approach. While it might be true that, to a degree, majority rule is about adjudicating fairly between competing, irreducible preferences, this question pertains to the procedural aspect of a defense of majority rule, with which an epistemic argument need not be concerned (although a full-blown justification for majority rule probably ought to). An epistemic approach need not be concerned about purely individual interests, except to the extent that they are material informing aggregated judgments. As such, interests will be present in a mediated

[8]I specify "random mechanism" to indicate that I am aware of the risk of agenda manipulation and to suggest one way out of this problem. Another option could be simply to ensure the political accountability of the agenda setter to the larger community (e.g., through election or peer pressure or some other mechanism) so as to suppress his or her possible partisan motivations. While some potentially epistemically damaging indeterminacy may result from the order in which the pairs are formed, this indeterminacy seems to be an unavoidable cost of decision making, whether the decision maker is a minority or a majority.

way in the final aggregated judgment, but they will not be the objects of the aggregation per se.[9]

This is not to say that the preference aggregation view of democracy is entirely irrelevant. But it is relevant only to the extent that politics is—as it is also but not only—about pure coordination games and pure conflict of interests or values. In such cases, when no better or worse solution for the group exists, the best thing to do is to take people's preferences (interests or values) and try to satisfy as many individual preferences as possible. A purely preference-based approach is valid, but only where no common good can or should be assumed, whether it is because the procedure will determine the common good (pure coordination games) or because we need to adjudicate fairly between antagonistic claims involving conflicts of interests or conflicts of values (on topics where we have agreed to disagree).

Trying to accommodate everyone's preference can also be worked out through negotiations and bargaining, which involve some amount of discussion and the search for a common agreement. It seems strange, however, to count this type of conversation as a variety of "deliberation," since these activities have the satisfaction of self-interests rather than the common good as a primary goal.[10] In such discussions, interests do not simply inform a position; rather, they can justify it. They do not work simply as a constraint on an argument but *are* an argument. In classical deliberation, by contrast, the interests at stake should be known but as such they do not *ipso facto* turn into justificatory claims.[11]

Regarding the case of pure conflict of interests, the domain for such situations will vary in size depending on how consensual the democratic community is. Jane Mansbridge (1980) defends the view that a democracy is always partially and more or less "unitary" and "adversarial" at the same time—that is, operating on the assumption of a common good and yet also divided by irreducible conflicts of interests. This is undeniably an accurate description of reality. The epistemic argument I am interested in, however, makes sense only for the "unitary" part of democratic decision making, the context where some common ground can be

[9]In denying that individual interests are directly relevant for the epistemic approach to democracy, I find myself opposed to the group of deliberative democrats that has recently tried to reintegrate interests, bargaining, and even power relations into a more comprehensive definition of "deliberation" (Mansbridge et al. 2010). See also chapter 4.

[10]Calling some of these negotiations "deliberative negotiations," as Mansbridge and her followers do (Mansbridge et al. 2010), only stresses the fact, in my view, that these practices fundamentally differ from what deliberation is about and only mimic the *form* of deliberative exchanges, not the substance.

[11]The exception is if the deliberating group finds that an interest is also in everyone else's interest, as the case of a woman protesting a highway construction on her land, when it can be shown that her land has historic value and should be preserved as a human heritage.

found between citizens and one can expect that they will go beyond their own self-interest and toward the good of the community as a whole.

I would extend that claim about individual interests versus common interests to values (as fundamental social norms) as well. In that regard, "the fact of disagreement" noted by many a contemporary liberal theorist—the fact that individuals in any given society differ fundamentally as to their conceptions of the good life and other fundamental values—should not be blown out of proportion. The normative importance of value pluralism as both a goal of liberal politics and one of the causes of this irreducible "fact of disagreement" characterizing liberal societies (Waldron 2001) does not raise an insurmountable challenge for the epistemic argument for democracy. It is possible that some conflicts of values stem from incommensurable worldviews. Those can probably be adjudicated only in the same way that pure conflicts of interests are—by a fair procedure.[12] But, to repeat, many functional democracies rest on a basis of shared values and some amount of consensus on a great number of issues. This basis of shared values and consensus is necessary for us to be able to speak of "a people" in the first place, as opposed to a multitude of atomistic individuals or a "balkanized" entity where groups coexist without a common sense of purpose and where political order itself is always in danger of collapsing into a civil war.

An epistemic approach, however, assumes a consensual background of shared values, which may range from very thick to very thin. Even arguably "pure" proceduralist accounts of deliberative democracy (e.g., the early Habermas, before the 2006 article) also must presuppose such a shared background (which Habermas calls "background understandings" of the lifeworld[13]). With such a background, which includes a commitment to democratic procedures, disagreement about facts, interests, and less fundamental values is not necessarily threatening. It signals that a debate is ongoing and that the quest for better solutions is a constant work in progress. The better solution on which everyone would agree may never be found, but it nonetheless remains a valuable and meaningful regulative idea. In that scenario, the "fact of disagreement" may merely stem from what Rawls (1993) would call, in constitutional matters, the "burdens of judgment," and what are in the end the information, time, and cognitive constraints under which citizens form their judgment.

---

[12] And it is true that in some cases—say, Lebanon, a country deeply divided on religious and cultural fronts—the degree of value pluralism is such that it probably makes it impossible for the epistemic properties of democracy to be expressed.

[13] Though these can always themselves become issues for deliberation and be contested, they nonetheless are the necessary backdrop for any deliberation at all. I owe this general remark on Habermas to Erin Pineda (Yale University).

2. The Problem of Informational Free Riding

Another type of objection to the epistemic properties of majority rule questions the assumption that voters in mass democracies are sufficiently informed to meet the minimal threshold of competence necessary for their aggregated judgment to have any epistemic value at all. The reader can go back to chapter 2 for a review of some of the literature on this question. Here, I will solely be interested in an important theoretical reason supporting the suspicion, if not the evidence, that voters in mass elections are uninformed: it is the rational choice theory assumption that citizens in mass elections have no incentives to become properly informed because their individual votes, being very unlikely to be pivotal, make virtually no difference to the outcome. Rational choice theory, on the basis of this assumption, predicts that voters will generally abstain from voting or, if they end up bothering to vote for some irrational reason, will vote in a profoundly uninformed manner. This objection can be generally described as the so-called informational free-riding problem, which arguably plagues large elections.

I will here assume that voters' levels of information, as measured by public opinion polls and contemporary political sciences, are a relevant factor in the production of epistemically valid collective outcomes; I thus postpone my discussion of the empirical claim that the public is abysmally misinformed, particularly compared with consumers, to the third section of this chapter when I address Bryan Caplan's recent version of this critique. I should here mention that the theoretical objection of informational free riding is distinct from Caplan's fear of systematically biased voters, another objection I deal with in section 3, in that the disinformation assumed here can take any form. The argument is, simply, that if all voters are all equally misinformed, as they should be given the incentive structure of large elections, they will vote randomly, and this, absent the further assumption of an informed elite of voters, cannot produce any smart prediction in the aggregate.

Here, one possible answer consists in rejecting the rational choice theory premise that voting makes no sense unless you are the pivotal voter, without however embracing the classical answer to that problem offered by advocates of an "expressive theory of voting" (e.g., Brennan and Lomasky 1993; Brennan and Hamlin 1998). Both critics of voting and its expressivist advocates share the view that citizens have no reason to become properly informed. Why should you make an effort to become informed if your vote makes no difference? Even assuming that voting is not pointless and that you do so for noninstrumental, expressive reasons ("cheering your team"), you still do not have a reason to vote in an informed manner.

Instead it can be argued that voting in an informed way is instrumentally rational, even in large elections. Richard Tuck (2008), for example, argues that the current social scientific manner of reasoning about the rationality of voting is profoundly contingent, as it is historically dependent on twentieth-century economic assumptions about perfect competition. In contrast, he contends, the commonsense view he embraces (in short, that it is generally instrumentally rational to contribute, in however a small way, to a large public good) existed prior to this paradigm shift in the social sciences and still exists outside of them. Presumably, in Tuck's view, if it is minimally rational to vote, then it is similarly rational to become informed, even in large elections.

Gerry Mackie provides a related though different defense of the rationality of voting, which also conveniently addresses the issue of informational free riding as a by-product and is rooted in what he calls the "contributory theory of voting" (Mackie 2012). This theory argues that voting has both an instrumental value (my vote causes my team to win when it is pivotal) and a mandate value (my vote causes my team to win by a certain amount or lose by a certain amount). According to this theory, the reason people vote is that they plausibly expect to contribute, in however negligible a way, to the margin by which their team is going to lose or win, and when multiplied by the large public good that could result from voting, this negligible contribution is often worth making to the voter. As the election is closer, the marginal benefit becomes more important since one vote might indeed be pivotal and the margin by which one's camp may lose becomes more crucial.

The contributory theory of voting combines the instrumental value and the mandate value of voting in such a way as to make voting rational always, but particularly when the election is close. For the same reason that each voter has a small but not insignificant reason to vote, she has a reason to become informed even in mass elections, particularly if those are close elections. As Mackie sees it, voting is not (one might add, *just*) a matter of cheering your team, but a matter of playing on the team. While it may not matter for the outcome of the game how fit the supporters are, it certainly makes a difference if the players are properly trained and in good physical and mental condition. Similarly, the contributory theory of voting implies that it does matter to the outcome of an election whether my individual vote is not just cast but informed. While the classical rational choice theory approach and the expressive theory of voting supposed to counter it concur that large elections are (in theory) threatened by rational informational free riding, the contributory theory of voting does not and, in effect, predicts that voters still have a small but nonnegligible reason to become properly informed. In effect, Mackie argues that voters are generally informed enough for the kind of voting expected of them in a representative democracy (Mackie 2012).

I will not pursue much further the refutation of the objection of informational free riding, in part because it presupposes that we have solved the question of what kind of information matters to good epistemic outcomes—a question I believe political sciences still have a relatively crude and misguided view on—and the further question of whether our current ways of measuring the information held by citizens are adequate. Let me turn instead to those issues in the next section.

3. The Problem of Voters' Systematic Biases and Their "Rational Irrationality"

The main problem with the optimistic conclusions about group intelligence that I have derived in the previous chapter is that in some way or another they rely on the assumption that there is a symmetrical distribution (random or otherwise) of errors around the right answer (Miracle of Aggregation) or that errors are negatively correlated (Hong and Page's account [2007, 2012]). But why should we not assume, on the contrary, that voters make highly positively correlated mistakes or have asymmetric biases? Here, I will rely on the case made by the economist Bryan Caplan (2007) that voters are worse than uninformed and that they are, in effect, systematically wrong, so that democracies are doomed to choose "wrong policies." In my view, this case does not refute the epistemic argument for majority rule and, more generally, democracy proposed in this book, although it certainly invites some caution in the definition of where democratic reason applies. The following, although specifically aimed at Caplan, has the broader goal of showing the limits of objections to democratic reason based on the denunciation of voters' systematic biases. I will raise a series of objections to the antidemocratic implications of Caplan's book, addressing in the process the larger question of the relationship between information and political epistemic competence.

The first basic reply that can be made to Caplan's critique of democracy is that empirical observations of the way American democracy functions or fails to function are not sufficient in themselves to falsify a more general claim about the epistemic properties of democracy as an ideal type (the type of claim I'm concerned with in this book). The empirical problems Caplan points out may be due to the fact that American democracy is not a real democracy in the sense used in this book, lacking too many of the features I have insisted on (e.g., a representative system preserving the cognitive diversity of the larger group). Caplan, however, backs up his empirical observations with a theory of the "rationally irrational" voter, which seems much more worrying. I will take seriously both the empirically and theoretically motivated objections and will address them in turn, starting with the theoretical objection.

3.1 The Rationally Irrational Voter

The theoretical objection is powerful if one accepts Caplan's general conceptualization of the voter. According to Caplan, voters have preferences over beliefs and maximize the ideological pleasure of feel-good beliefs. Taking into account the lessons of psychology and behavioral economics, Caplan argues that citizens' judgment is afflicted by systematic cognitive biases such as "the antimarket bias" or the inability to understand that private greed can benefit the public good.[14] As a result of such biases, citizens cannot be expected to want the means to their own preferred ends. Thus, to get the people what they want, the last thing one should do is to ask them for their opinion. In effect, Caplan argues, we should have much less democratic input, particularly on economic issues. In other words, we should let economists rule, or better still, the market.[15]

There are, however, many problems with this conceptualization (see also Elster and Landemore 2008 for a critique). In particular, Caplan assumes in voters a form of self-interest incompatible with the framework of this book (and empirical work supporting the view of voters as genuinely altruistic and other regarding, e.g., Bowles and Gintis 2006). Even when they vote ideologically (for the "common good") as opposed to "rationally" (for their pocketbook), as in for example, the case of the rich Hollywood actor voting for higher taxes, voters are only doing so, on Caplan's account, because of the unlikely prospect that their vote would be pivotal. So, in effect, the voters are still first and foremost preoccupied with voting their self-interest (in the form of a warm-glow effect when the impact of their vote is too low to make a difference) rather than promoting something like the common good.[16] Arguably, on Caplan's account, if their vote mattered at all, they would revert to voting their pocketbook. This perspective is utterly incompatible with the epistemic framework of the argument presented in this book, which assumes that people are voting

---

[14] Other systematic cognitive biases denounced by Caplan are what he calls the antiforeigner bias (a distrust of foreigners that leads to overly protectionist policies), the underestimation of the benefits of conserving labor, and the pessimistic bias (the mistaken belief that the economy will go from bad to worse).

[15] See section 2.1.1 of chapter 2 in this book.

[16] There is in fact an incoherence in the description of the "rational purchase of altruism" (see Elster and Landemore 2008).

what they think is right for the common good, no matter how unpleasant it is for them, whether ideologically or economically. The theoretical divergence runs so deep that it is hard to see any point of intersection between those two models, which lead to drastically different predictions about the epistemic quality of democratic output.

Regardless of that theoretical divergence, however, what about the objection that even an epistemic framework may be challenged by the existence of systematic biases in voters, whether these biases come from ignorance, irrationality, or anything else? It is true, as already noted in the previous chapter, that an account of collective intelligence based on cognitive diversity is no more immune to the problem of systematic biases than the Condorcet Jury Theorem or the Miracle of Aggregation are. If citizens share a number of wrong views—racist prejudices or systematic economic biases—majority rule is simply going to amplify these mistakes and make democratic decisions worse, if anything, than the decisions that could have been reached by a randomly chosen citizen. In the account of collective intelligence that I embrace, however, which emphasizes cognitive independence, the risk of systematic mistakes can only happen if the group lacks both individual predictive accuracy (i.e., people are not sufficiently intelligent) *and* diversity in the way they make predictions. Assuming minimally sophisticated voters relative to the questions at hand and a liberal society encouraging dissent and diverse thinking, however, Caplan's worst-case scenario of a situation in which the average error is high and diversity low—the conditions for the worst-case scenario of an abysmally unintelligent majority decision—is not very plausible. In other words, the possibility of systematic biases on a majority of political issues is not very plausible.

Furthermore, deliberation can play a role in the epistemic argument for democracy developed so far that it cannot play in Caplan's model. When it comes to majorities making mistakes, my argument at least allows for the possibility of self-correction over time and through the means of public deliberation, whereas Caplan, it seems, would either bring in the experts or exit politics altogether (in favor of the market). In actual democracies, it is interesting to see that where systematic bias scenarios have been historically observed to exist—on race issues, for example—most changes had to come from evolving majorities themselves, through a democratic process of collective self-reflection and public deliberation.[17] Democratic deliberation, which includes the experts as

---

[17] In that deliberation, some may want to see key Supreme Court decisions as a part of, and some others as an alternative to, the democratic dialogue, which somewhat complicates the equation. But one could always argue that the way constitutions tie the hands of the people on some issues was itself an initially democratic decision that the people made at some point in order to protect themselves against their own predictable irrationality, by creating constitutional safeguards for minorities, for example (Elster 2000).

welcome but nonexclusive voices, is a central part of the argument made in this book and offers a possible solution to the problem of the occasional systematic mistakes that the public can make, a solution never seriously entertained by Caplan.

3.2 The Objection from the Empirical Evidence of Systematic Cognitive Biases

Let me now turn to the empirical challenge based on the measurement of systematic biases in actual American democratic citizens.

Using empirical evidence borrowed from the literature on "enlightened preferences" (essentially Althaus 2003), the Survey of Americans and Economists on the Economy (SAEE),[18] and the results of his own comparison between the public's preferences and those of an "enlightened public" virtually endowed with a PhD in economics, Caplan diagnoses four main misconceptions held by the average American citizen with respect to economic questions: an antimarket bias, a protectionist bias, a pessimistic bias, and a job-oriented bias. Assuming that economists are right that all things being equal otherwise, the market mechanism is a good thing, free trade creates more riches than it destroys, growth is more likely than stagnation, and GDP increase matters more than job preservation, then the people are wrong to hold opposite views and ask for policies based on such beliefs. The problem is not solved, or solved only to an insufficient degree, by the fact that policies are made a priori by slightly more competent representatives. To the extent that representatives are held accountable to the citizens, they have only limited leeway to improve the course of things. Consequently, Caplan concludes that, on economic questions at least, we would be better off with less democratic input. He himself seems to suggest more delegation to economists and, whenever possible, to markets themselves.

Here, I will raise three criticisms, each of which are elaborated in more detail below. The first criticism bears on the elitist premises of the book and the method used to measure citizens' incompetence. There are at least four different standards in the book, serving as benchmarks of citizens' biases: objective facts with a verifiable answer, the simulated "enlightened preferences" of a public with high political IQ, the simulated preferences of an "enlightened public" with the knowledge of a PhD in economics, and finally the policy preferences of economists themselves. The problem

---

[18]The survey is based on interviews with 1,510 randomly selected members of the American public and 250 economic PhDs and designed to test for systematic lay-expert belief differences by asking questions such as whether various factors are "a major reason," "a minor reason," or "no reason at all" for why "the economy is not doing better than it is."

is essentially that (a) objective facts are not a conclusive standard (the relationship between the possession of factual knowledge and epistemic competence being too shaky); (b) taking economists' knowledge as the standard begs the question of who has authority in politics in the first place; and (c) the other two—"enlightened preferences" or "enlightened public"—are in fact slight variations about either facts or expert knowledge.

Second, even granting that Caplan is right about the economic incompetence of the average voter, the implications for democracy are not nearly as bad as Caplan would like to suggest. Finally, I object to the alternatives implicitly offered by Caplan. It is indeed unclear that the oligarchy of experts that Caplan sometimes seems to advocate would necessarily do much better overall than a democracy. As for a market mechanism, it is not a political alternative to any form of government but a mere allocation tool in the hands of the one, the few, or the many—thus leaving untouched the question of who should rule.

Let me first address the methodological question. The first benchmark of voters' bias is knowledge of objective facts. As Caplan observes, "the simplest way to test for voter bias is to ask questions with objective quantitative answers, like the share of the federal budget dedicated to national defense or Social Security" (Caplan 2007: 25). Caplan, however, does not dwell on that first standard, acknowledging that "the main drawback of these studies [that measure the mastery of factual knowledge] is that many interesting questions are only answerable with a degree of ambiguity" (ibid.). Indeed one could argue that *no interesting political question* can be answered without such a degree of ambiguity, which raises the general issue of the relevance of a great deal of public opinion research that measures the ability to answer textbook political questions. Since the standard of objective facts reappears through the back door of the notion of "high political IQ individuals," let me say a few more words about why this standard is unsatisfying.

Information is distinct from competence, and the causal link between the holding of a certain type of information measured by surveys and the competence to make political choices is not easy to establish (however "intuitive" it is sometimes argued to be). In fact, most existing studies (e.g., Luskin 1987; Delli Carpini and Keeter 1996) fail to demonstrate a causal link between the inability of people to answer certain types of political quizzes and their alleged political incompetence, namely the inability to make the right choices or hold the "right" policy preferences. This is in part because the design of factual political questionnaires smacks of elitism, measuring a type of knowledge relevant for policy analysts and journalists, but not necessarily the only one conducive to smart political choices (Lupia 2006).

The difficulty of establishing a causal link between low information level and political competence comes also from the lack of a good empirical benchmark for politcal competence that would be distinct from a good benchmark for information level. The fact that educated people are good at answering political quizzes does not entail (1) that the policy preferences of the educated are better as a result (unless you take such policy preferences as the standard, but then you are begging the question) or (2) that the policy preferences of "know-nothings" or low-political-IQ individuals (as defined by such tests) are wrong.[19] The kind of factual knowledge measured by public opinion surveys is a crude measurement of political competence, and there is no reason why the burden of the proof should be on people who deny the connection between political IQ as it is measured by existing empirical surveys and actual political competence.

Let us now turn to the second standard: the "enlightened preferences" of a hypothetical educated public—a group of people that is demographically representative, except that they are as politically knowledgeable as possible. The method used by Althaus (2003) consists in administering a survey of policy preferences combined with a test of objective political knowledge to a group, estimating individuals' policy preferences as a function of their objective political knowledge on factual matters (e.g., how many senators each state has) and their demographics (income, race, gender), and, finally, simulating what policy preferences would look like if all members of all demographic groups had the maximum level of objective political knowledge. In other words, the goal is to compare the policy preferences of regular citizens with those of their "more educated" counterparts, controlling for race, gender, income, and the like.

The enlightened preference approach permits testing of the plausibility of the theory of the "reasoning voter," according to which the votes of people with little knowledge are roughly the same as if they had maximal information thanks to cognitive shortcuts, heuristics, and online processing. The major result of this approach is to show that, no, people would probably not vote the same way, based on the fact that the relatively uninformed voter does not have the same preferences as her very informed counterpart: voters tend to be more socially liberal and economically conservative in the second case (Althaus 2003: 143). Scott Althaus uses the discrepancy between the public's preferences and those of its more "enlightened" self to criticize the representativeness of opinion surveys

---

[19] After all, even the writers of the TV show *The West Wing* knew that you can be a competent director of communications at the White House and be unable to say three correct things about the history of the White House (*West Wing*, episode 1).

and their usefulness in assessing the public's voice. Caplan goes one step further, using those results to suggest that democracy itself, which follows the more or less unenlightened policy preferences of the many, is flawed.

Consider however that the definition of "enlightened preferences" hinges on a concept of education that is correlated with the ability to score well on political IQ tests ("a test of objective political knowledge"). The standard of "enlightened preferences" is thus not much different from the knowledge of objective facts (since it is highly correlated with it). But we just saw that knowledge of objective facts might well be both an elitist measure of political knowledge and potentially irrelevant to the ability to vote in a politically competent way. So what this approach does is take as the standard of "enlightened" judgments preferences correlated with an elitist and possibly irrelevant form of knowledge and then argue that the discrepancy between the actual public's preferences and those "enlightened preferences" is not only meaningful but, in fact, an embarrassment for democracy. Such conclusions, however, merely reflect a belief present in the premise, namely that regular people are wrong and the elites are right. How is that not begging the question of who has epistemic authority in the first place?

The third standard consists of the economic preferences of a simulated public that is both demographically representative and endowed with the knowledge of a holder of a PhD in economics. The key difference between Caplan's approach and the previous approach is that "political scientists usually measure knowledge directly, while my approach proxies it using educational credentials" (Caplan 2007: 55). So, in effect, whereas the second type of approach boils down more or less to using the standard of objective facts (through the notion of political IQ) to assess the public's preferences, Caplan's approach takes as the ultimate standard the knowledge of experts. Another difference is that the competence that Caplan is trying to assess is slightly narrower than that measured by political scientists, since Caplan is interested only in political questions with an explicitly economic dimension, for which economic knowledge such as that measured by a PhD diploma might seem directly relevant (more so, at least, than "objective political knowledge" with respect to political competence).

So let us consider why this, and the fourth standard—experts' knowledge—is problematic. First, Caplan constantly writes as if there was no difference between questions of economics (the science) and economic questions, which are political questions with an economic dimension. Just because PhD holders in economics are the best at answering questions in the science of economics does not make them the most competent at answering political questions with an economic component (although their input is most likely of value). In fact, if you deny that economists' political beliefs are absolute truths, the discrepancy between these beliefs and those of the public does not necessarily say much.

Despite initially acknowledging that political questions cannot be answered without a degree of ambiguity, Caplan does write as if the beliefs of economists were on a par with mathematical truths. Here is a typical example. Caplan argues that "elitist though it sounds, [inferring the existence of systematic biases in the public from the existence of systematic differences between economists and noneconomists] is the standard practice in the broader literature on biases" (Caplan 2007: 52). Caplan goes on to appeal to the authority of no less than Kahneman and Tversky, who describe their own method this way: "The presence of an error of judgment is demonstrated by comparing people's responses either with an *established fact . . .* or *with an accepted rule of arithmetic, logic, or statistics*" (cited in ibid.).[20] Caplan thus draws a clear parallel between the consensual beliefs of economists, on the one hand, and objective facts or the rules of arithmetic, logic, or statistics, on the other hand.

This parallel, however, is highly misleading. To the extent that economic beliefs are about facts (the share of foreign aid in the federal budget) or about mathematical theorems, they are not necessarily relevant, or not directly so, for political decisions. To the extent that these beliefs are more "political"—even the least controversial ones, like "free trade is good" or "people are not saving enough"—they are much more contingent on a shifting cultural and possibly ideological consensus among experts than Caplan allows for. By playing on this ambiguity between pure questions of textbook economics and political questions with an economic dimension, and by misleadingly identifying the beliefs of economists at a given time with factual truth or mathematical principles, Caplan indeed begs the question of who has authority in the first place. In his view, on anything remotely economical, economists know better. If you deny that premise however, none of Caplan's conclusions follow.

Both the "enlightened preference" approach and Caplan's "enlightened public" approach implicitly raise the question of who is politically competent in the first place, whether it is people with a high political IQ or economists. Caplan supports that way of proceeding by arguing that "the burden of the proof should be on those who doubt the common sense assumption that we should trust the experts" (p. 82). One might reply, however, that democracy is premised on the very rejection of that

[20] Caplan further comments "'established' or 'accepted' by whom? By experts of course" (p. 52). Notice, however, that unlike mathematical truths—which can be accepted by everyone, not just experts—economic truths are never as universally endorsed.

"commonsense" assumption. Recall the assumption laid out in chapter 1 about the nature of politics as the space where communities deal with the unknown and the uncertain. Recall also, from chapter 3, how the profoundly democratic practice of *isegoria* could be interpreted as stemming from a belief that in politics specifically, by contrast with more technical domains, all voices ought to be heard because no one has a privileged claim to knowledge over anyone else.[21] For Athenian democrats, the real test of competence and expertise in politics is thus the ability to convince others in the assembly. This is why, even if ultimately only the better arguments and information are supposed to triumph, everyone has the right to speak up.

I just criticized Caplan for begging the question of who is right in politics when defining the benchmark of competent answers as those of people who think like economists. Caplan, however, might simply retort: Aren't democrats begging the question the other way around by denying that there are experts in the first place?

The positions are not exactly symmetrical. In Caplan's case, the question of who knows best and what the right answers are is a priori locked and determined. The economists know better—their answers are the right ones—and thus any deviation from their position must be measured as a bias. In the democratic view, in contrast, there is genuine agnosticism as to who knows best and what the right answer is, at least at the outset. Who knows best and what the right answers are can be determined only on the merits of different claims competing in public space. Of course, just like economists' claims, the merits of such competing claims are only firmly established retrospectively, by judging how well the country did overall given that such and such policies were implemented or even by comparing expected to actual results for every chosen policy. The telling difference, then, is that at the moment of decision making, when such hindsight is not available, the benchmark of right political answers is, for Caplan, whatever economists say, whereas for democrats the benchmark is only the "force of the better argument" (which does not mean that the best argument will always triumph) and/or majority outcomes (which does not mean that the majority is always right).

Of course, when looking at the actual discrepancy between what the public thinks and what economists think, there are cases in which it may seem like the experts are probably right and the public probably wrong. For example, people tend to think that "taxes are too high" or that "foreign aid spending is too high" (Caplan 2007: 57), whereas economists and the "enlightened public" sensibly differ. A lot hinges on what is meant

---

[21] According to Cynthia Farrar (1988), "Protagoras was, so far as we know, the first democratic political theorist in the history of the world" (p. 77).

by "too high" and with respect to what benchmark these assertions are made. But I concede that it is possible, on some questions, that there are such biases, in which case the superiority of an oligarchy of knowers who would be able to avoid those mistakes would be established. There are, however, a few reasons why even granting occasional topical incompetence does not affect the general argument for democracy developed in this book.

First, topical incompetence does not establish global incompetence and, in particular, the meta-incompetence to recognize one's topical incompetence. Even if we accept that citizens are bad at answering political questions of an economic nature, that does not mean that they are not reasonable enough—that is, minimally competent—to acknowledge that fact and accept institutional arrangements that compensate for it, such as delegation of some decisions to acknowledged experts.

Second, delegation of some choices to experts does not imply the failure of democracy. Democracies that delegate some decisions to a few unelected individuals do not ipso facto turn into oligarchies. The fact that the consent of the people is initially obtained for this delegation to take place (directly or through their representatives) still makes the decisions of those experts "democratic" in a larger sense. To the extent that the independence of central banks itself is a democratic choice, it should testify to democratic intelligence on Caplan's view, since his story is supposedly voter driven. Conversely, the decision power of democratically authorized experts on some economic questions does not prove the superiority of oligarchy over democracy but simply establishes the necessity of having some efficient technocratic cogs in a larger and more complex democratic structure of governance.[22] The relevant comparison for my purposes in this book is not between democracy and that technocratic branch of the government but between democracy and oligarchy when both are equipped with a competent technocracy of that kind. John Stuart Mill thought that the only virtue of a monarchy was its bureaucracy, whereas the virtue of a democracy was its bureaucracy plus the intelligence that goes into overseeing it (Mill [1861] 2010: chaps. 5 and 6). Similarly, my argument leads me to conclude that when both democracy and oligarchy are equipped with a competent army of experts, democracy should still, on average and in the long term, outperform oligarchy.

Third, even if Caplan is right about voters' topical incompetence, particularly in economic matters, why not consider the possibility that such topical incompetence might be solved over time through education and public debates? I already mentioned that deliberation might be a solution

---

[22] In fact, this voluntary delegation of technical economic questions to experts is all that Caplan should ultimately advocate.

to systematic biases. But Caplan seems to equate observed ideological preferences with deeply entrenched (bad) cognitive biases and heuristics. In the same way that people are known to suffer from base-rate neglect[23] or to be subject to framing effects,[24] Caplan suggests that they are systematically anti–free trade and pro–job security. But an anti-market or a pro-job bias is of a different nature than an inability to calculate probabilities correctly or see a glass as equally half full and half empty. Such economic biases are less due to the limits of human cognitive abilities and more to cultural factors. After all, while all human beings may suffer from some form of base-rate neglect, Americans are actually much less obsessed with job security than Europeans. President Clinton during his presidential campaign could thus warn the American public that "they would have to change jobs seven to eight times in a lifetime"—a discourse utterly unthinkable in a French context. Racial and sexist prejudices have considerably diminished in most Western democracies over just a few generations.[25] These facts suggest that some biases can be corrected, at least partially. Maybe economic biases are of a more enduring nature, but Caplan does not demonstrate this for a fact. Education and a more deliberative democracy, however trite that may sound, may well be the answers to (at least some of) the flaws of our existing democracy.

The final objection I will raise is against an apparent implication of Caplan's indictment of democracy—that we would be better off with an oligarchy of experts[26]—is that groups of experts are not foolproof, either. Philip Tetlock (2005) showed in his study of "political judgment" that when it comes to assessing a problem and making political predictions, political "experts" do hardly better than laypeople and, on the purely predictive side, are in general outperformed by simple statistical regressions. Striking what should seem like a deadly blow against the idea that politics is a matter of expertise, Tetlock concludes that it does not really matter *who* the experts are (economists or political scientists or philosophers, etc.) or *what* they think (ideologically, e.g., whether they tend to be pro-market or socialist). What matters is rather

---

[23]The base-rate neglect or fallacy consists in neglecting the prior probability of some hypothesis H when trying to assess the conditional probability of this hypothesis given some evidence E.

[24]They give different answers to the same question framed differently.

[25]The United States has now a black president.

[26]Caplan would deny that this is the solution he advocates, yet everything, from the cover of the book to many assertions in it, invites an antidemocratic reading. Caplan could have tried harder to dissuade the reader from thinking that what he ultimately advocates is rule of the experts, in the same way as he would have liked to see Tetlock be clearer about the fact that his book, according to Caplan, does not establish the superiority of the layman over the expert (Caplan 2007).

the *way* political experts think, namely whether they think as "foxes" or as "hedgehogs."

Borrowing Isaiah Berlin's ideal types, Tetlock characterizes foxes as eclectic thinkers with an ability to use different frameworks and theories. By contrast, hedgehogs are dogmatic thinkers with a one-size-fits-all theory of the world. From what Tetlock could empirically observe, foxes are almost always better forecasters than hedgehogs. Tetlock also shows that both foxes and hedgehogs are generally outperformed by statistical regressions. If political experts—pundits, political campaign leaders, diplomats, and so on—tend to overestimate their knowledge, analytical skills, and ability to predict what will happen, it is probable that economists—who tend to fit the model of the hedgehog, or dogmatic thinker described by Tetlock—suffer from the same cognitive failures.

Do Tetlock's results imply that there is no added value to expert advice as compared with the judgment of well-informed laities? Concludes Tetlock: "In this age of academic hyperspecialization, there is no reason for supposing that contributors to top journals—distinguished political scientists, area study specialists, economists, and so on—are any better than journalists or attentive readers of the *New York Times* in 'reading' emerging situations" (2005: 223). In reply to this, Caplan argues that one should not misinterpret the meaning of Tetlock's results. According to him, all that Tetlock shows is that experts are bad at answering difficult questions, not easy ones, which does not imply that laypeople would do much better on either type of questions (Caplan 2007). Fair enough, but that still not does give us a decisive argument why we should ultimately trust economist experts more than laypeople (or their representatives) when it comes to making political decisions, including when those decisions have an economic component. In fact, the argument from diversity presented in earlier chapters implies that lack of cognitive diversity among experts can offset the advantage represented by their individual expertise, while, on the contrary, the cognitive diversity of large groups of nonexperts can compensate (to a degree) for their lack of individual expertise. In terms of predictive accuracy, large groups of laypeople and small groups of experts may well draw a tie.

CONCLUSION

Many objections can be raised against the idea that majorities can be smart. This chapter has addressed four main concerns: (1) various problems raised by social choice theory and, more generally, the view that politics is about preference rather than judgment aggregation; (2) the informational free-riding problem that arguably causes individual voters to be poorly informed; (3) the problem of "rational irrationality"; and (4) the related claim that voters are bound to be systematically biased. While all these objections are serious and worthy of consideration, I have tried to show why they need not form an insurmountable case against majoritarian decision making, particularly when the latter is preceded by appropriate public deliberation and is placed in the context of a functioning representative democracy. For the occasional and serious lapses, however, the procedures of democratic reason must accommodate the lessons of past mistakes, whether those are embodied in counter-majoritarian mechanisms—such as, notoriously, constitutional courts and the principle of judicial review—or other mechanisms. I return briefly to this issue in the conclusion of the book.

CHAPTER EIGHT

Political Cognitivism: A Defense

AN EPISTEMIC ARGUMENT for democracy depends on the view that at least for some political questions there are right or correct answers (in some sense of right or correct that remains to be defined) and that these answers can be, if not known with certainty, at least approximated to some degree. I label "political cognitivism" the combination of the assumption that there exists such a standard and the belief that it can be approximated in some way by a political decision mechanism.

Political cognitivism must make sense for at least some domain of politics in order for the cognitive argument for democracy to even get off the ground. This chapter lays out the argument for this view. The chapter also goes back to the question of the domain of validity of the epistemic argument (already touched on in the introduction), addressing the question of whether it includes facts, values, or both. The goal is to achieve a fuller picture of the epistemic case for democracy and what it would take to endorse it. While the chapter surveys a number of versions of political cognitivism, it does not specifically argue for any. Although I will make it clear where my own sympathies lie, the point is not to single out one option or interpretation out of the many that are available but, on the contrary, to show that political cognitivism is a position that can be embraced from a great number of perspectives and is compatible with a number of metaethical views.

The ambition of this chapter is limited. It is simply to render palatable to the skeptical reader the view that it makes sense to presuppose and aim for a procedure-independent standard of correctness, whether we decide to call it truth, rightness or otherwise, when we engage in political deliberations or when we vote (see also Estlund 2008; Goldman 1999, especially chap. 7 on democracy; Raz 1990; and Talisse 2009). I will not, however, explore in all its epistemological and metaphysical depth the question of the possible meaning of truth in politics, nor will I attempt to distinguish it from its meaning in morals, science, or other fields. For such an investigation, the reader is invited to consult the literature in social epistemology, particularly the debates between "classical" or social epistemologists, who focus on the epistemic goal of having true (or at least justified) beliefs, and "anticlassical" or postmodernist social epistemologists, who have no use for the concepts of truth and justification (e.g., Fuller 1987, 1988, 1999; Goldman 1986, 1987, 1999; Kitcher 1990, 1993, 2001).[27]

The first section clarifies the meaning of political cognitivism by using Rawls's idea of "imperfect procedural justice." Just as imperfect procedural justice defines a framework in which justice consists in trying to approximate with imperfect human means, such as juries, an independently existing standard such as guilt or innocence, political cognitivism is a framework in which imperfect democratic decision procedures are used to approximate some independently given political outcomes or ideals.

The second section introduces a distinction between weak and strong political cognitivism. While "weak" political cognitivism defines the standard of correctness as the avoidance of major harms, "strong" political cognitivism posits a more substantive range of outcomes.

The third section then turns to the three idealized components of political questions: facts, fact-sensitive normative principles or values, and basic normative principles or values. I argue that it is possible to posit a standard of correctness for both facts and fact-sensitive values. Whether one also posits a standard of correctness at the level of basic values marks the difference between culturalist and absolutist political cognitivism, a distinction I turn to in section 4. Depending on whether the standard is defined as relative to a given culture or valid across time and space, political cognitivism will be more or less compatible with different metaethical views.

Section 5 considers the implications of these different forms of cognitivism in terms of the strength and plausibility of the epistemic claim for democracy.

Section 6 presents the results of an attempt to quantify a good political judgment and illustrates what standards of correctness would be on a culturalist view of political cognitivism.

Finally, section 7 turns to the "antiauthoritarian" objection to political cognitivism, according to which the idea of an independent standard of correctness dangerously smuggles in a theory of truth. Since truth and politics have proven to be a historically explosive mix, the objection argues that an epistemic approach to democracy is bound to have illiberal and possibly authoritarian implications. I consider both Hannah Arendt and Rawls as representatives of that worry about the role of truth in politics and show why their arguments do not amount to a refutation of the validity of political cognitivism.

---

[27]This division corresponds to the difference between *Synthese* and *Episteme: A Journal of Social Epistemology*, on the one hand, and *Social Epistemology: A Journal of Knowledge, Culture, and Policy*, on the other,.

1. POLITICAL DECISION MAKING AS IMPERFECT PROCEDURAL JUSTICE

Political cognitivism is, loosely phrased, the position according to which at least some political questions admit of something like a "better" or "worse" answer. Saying that a political question admits of a better or worse answer is equivalent to saying that there is a standard of correctness by which one can assess the result of the decision procedure used to answer that question. The standard is thus independent from the decision procedure. The idea is that the outcome of some political decisions—whether achieved through democratic means such as deliberation or majority rule or through nondemocratic procedures such as relying on the judgment of nonelected experts—is ideally assessed using a standard distinct from the procedure itself. This avoids tautological definitions of "correctness" such as, for example, whatever a deliberating group, a majority, or an expert decides (see also Coleman and Ferejohn 1986 and Cohen 1986 for a similar distinction between the interpretation of populism as mere majoritarianism, where the outcome of majority rule defines the general will, as in Riker 1982, and "epistemic populism," where majority rule outcomes are, at best, evidentiary of the general will).

Independence from the procedure does not imply independence from everything else. The standard might be dependent, for example, on a given context, history, or culture, or on the defining traits of human nature. Of course, one might also posit a procedure-independent standard that is also independent from anything else and functions as a kind of Platonic absolute or a "view from nowhere."[28] This approach would tie political cognitivism to an especially strong form of moral realism, the view that values have an existence independent of human subjectivity. Moral realism, however, is by no means required by the idea of an independent standard of correctness. This standard need not be a norm with a definite content but can refer to a norm-generating ideal procedure like Habermas's ideal speech situation or Rawls's original position.

Where it applies, political cognitivism requires us to conceive of political decision procedures in the same way as Rawls conceives of "imperfect procedural justice." Let me briefly develop this parallel, which is enlightening.[29]

---

[28] To borrow Nagel's (1986) title.

[29] Beitz (1989) similarly acknowledges that "the question of political justice seems more similar to that of defining fair rules of criminal procedure than to that of describing the conditions of a fair bet; it appears to represent a case of imperfect procedural justice in which the independent standard is a criterion for evaluating the substantive outcomes of the political process" (p. 47).

For Rawls, perfect procedural justice has two characteristics. First, it features an independent criterion for what constitutes the right kind of outcome of the procedure, namely an outcome that is fair or just. In that regard, perfect procedural justice must be contrasted to "pure procedural justice," which describes situations in which there is no criterion for what constitutes a just outcome other than the procedure itself (as in gambling, where the just outcome is whatever the dice dictate).

The second characteristic feature of perfect procedural justice is that it defines a procedure that guarantees that the fair outcome will be achieved. Rawls give as an example of perfect procedural justice the procedure that consists in having a cake cut by the person who will be last allowed to pick her slice, thus ensuring equal shares for everyone else (Rawls assumes that the cake cutter will want to maximize the size of her own expected slice).

Imperfect procedural justice, on the other hand, shares the first characteristic of perfect procedural justice—there is an independent criterion for a fair outcome—but applies when there is no method that guarantees that the fair outcome will be achieved. Imperfect procedural justice is illustrated by the case of a jury. Even though one assumes that there is a standard of justice independent of the procedure—the actual guilt or innocence of the person on trial—one can never be sure that the right answer has been reached.

Political cognitivism implies that political decision-making procedures resemble the procedures used to approximate a just order, only applied to a larger domain. Whereas imperfect procedural justice, at least as Rawls understands it, applies only in the realm of basic distributive questions about the basic structure of society, the epistemic approach to democracy posits a wider domain where there exists both an independent standard of correctness and a procedure, albeit an imperfect one, that offers us a reliable way of approximating this standard.

2. POLITICAL COGNITIVISM: WEAK VERSUS STRONG

Let me now try to make this idea of a procedure-independent standard of correctness less abstract. Is it possible to make mistakes in politics? Surely the answer is yes. Some political decisions turn out to be utter fiascos, about which we can say that almost every citizen would have been better off if they had not been made. Though many mistakes go unacknowledged, there are a few examples of political judgments in past or recent history that are now generally described as famous political mistakes: the Sicily expedition, European democracies' appeasement strategy toward

Hitler, the Bay of Pigs disaster, some aspects of the 2003 invasion of Iraq,[30] or simply famines that could have been avoided.

Policy decisions that result in military defeat or numerous deaths are probably the most obvious mistakes. But one may also classify as mistakes decisions that end up, for intrinsic rather than contingent reasons, disrupting the economy or the social fabric. "Shock therapy" policies, that is, the mix of wild liberalization and financial austerity implemented in a number of Eastern and Central European countries, have notoriously precipitated an economic and social situation sometimes worse for their populations than what they had known under the communist era. World Bank experts themselves generally admit that their recommendations, in these particular cases, did not always bear the expected fruit. In a different and perhaps less controversial vein, the French immigration and urban policies of the 1950s and '60s are now held largely responsible, by both the French Left and Right, as well as many external observers, for the explosive situation in the housing projects of French major cities, where racial ghettoization, underemployment, and general insecurity culminated in the highly publicized riots of fall 2005.

Of course, the most stubborn "hedgehogs" (Tetlock 2005) might object that the payoffs of, for example, neoliberal policies like "shock therapy" might lie in the future, so that policies that now seem wrong will seem correct in the longer run. This objection—"It's too soon to tell" or, in Tetlock's list of the many psychological defense mechanisms human beings resort to when their belief systems are challenged, the "just off on timing" argument—does not entirely cut it. Even beneficial austerity cures need to preserve a feeling of optimism and trust in the population that they supposedly help, short of which they risk being self-defeating in the long run too, like a diet so extreme that it would kill the person before she had a chance to even start slimming down.

A first way to look at the standard of correctness is thus as a thin standard of harm avoidance. According to this view, political cognitivism does not imply that all political questions have a right or wrong answer, that political questions with such answers have universally right or wrong answers, or that their answers are unique where they exist. It does imply, however, that better and worse decisions can be made. The independent standard need not involve definite, positive truths. But it requires that there is a realm of things that are the opposite of political mistakes. Thus, on a minimal or thin reading of the independent

---

[30] In the United States, while Democrats and Republicans may still disagree about whether going to war in Iraq was a good or a bad idea, they do agree that sending too few troops was a mistake.

standard of correctness, all that political cognitivism requires is the idea that political judgment, democratic or undemocratic, can err and that some decisions are, in some sense, wrong—and thus that others may be, in some sense, right. I propose to call the kind of political cognitivism premised on such a thin standard "weak political cognitivism." On a weak interpretation of political cognitivism, a political decision procedure that proves good enough at avoiding major mistakes can be said to produce "correct" results. James Bohman (1998), for example, defines the goal of deliberation not as "truth tracking" but as "error avoidance." In my view, this distinction is merely a distinction between strong and weak political cognitivism.

One may want, however, to posit a thicker, more substantive standard of correctness, defining some outcomes or a range of outcomes as the standard. For example, a classical utilitarian would define the standard as the maximization of welfare for the greater number. The utilitarian would not label a political decision as correct if it only avoided catastrophe without meeting this positive standard. A narrower standard to assess the validity of a decision could be, for example, whether it contributes to economic growth or to diminishing the mortality rate of newborns. Obviously, the standard of correctness will vary according to the question at hand, although some ends may be considered universally valid. It will help here to introduce a distinction between the factual components and the value components of political judgments.

3. The Three Sides of Political Questions

It has been convincingly argued that there is no such thing as a clear-cut "fact/value dichotomy" and that both the factual and the value aspects in most interesting questions are irremediably entangled, in the same way that a grey fabric contains indiscernible threads of black and white (Putnam 2003).[31]

Although I generally subscribe to this view, the distinction between fact and value remains analytically useful. It allows us in particular to distinguish between three aspects of political questions: facts, on the one hand, and two types of values, which I will call respectively context-dependent principles and basic values, on the other. These distinctions are meant as ideal types, as in reality all three elements are intertwined. We can distinguish a specific standard of correctness or truth for each of them.

---

[31] I am adapting Quine's famous metaphor and simplifying Putnam's use of it.

3.1 Facts

A first category of political questions is the factual type. Questions such as "Are there weapons of mass destruction in Iraq?" or "What was the rate of unemployment at the end of 2006?" may raise disagreement among citizens and experts alike, but we assume that they ultimately have a verifiable answer. "Verifiable" is used here in the most intuitive sense, referring to the possibility of checking the reality of the corresponding facts, more or less directly, by our senses (or by our senses aided by instruments). Other political facts are of a more theoretical nature, such as "Does the development of nuclear weapons by nondemocratic countries threaten the national security of the United States and world peace in general?" or "Does an increase in statutory minimum wages increase unemployment?" or "Does legal prohibition of drugs encourage or limit their consumption?" These questions involve not only facts difficult to define and measure with precision, such as national security or unemployment, but causal relationships as well, which are facts of a very immaterial nature. One may argue, however, that even such facts can be, ideally, sooner or later, or in some way, verified and checked against reality.[32]

Finally, among political facts is the category of historical events, whose truth can be said to be similarly anchored in perceptible reality. As Hannah Arendt reminds us, however, such historical facts are more fragile, as they are entirely dependent on societies' memory. Arendt thus gives the example of the fact of Trotsky's participation in the Russian Revolution, which was erased from Soviet history books (Arendt 1993: 231). She suggests that even such an apparently uncontroversial fact as "Germany invaded Belgium in August 1914" could potentially be written out of history if enough power was gathered in the wrong hands. Compared with what Arendt calls "rational truth" (the mathematical and scientific truths, which, according to Arendt, could theoretically always be "found again"), there is a contingency about factual truth, particularly of the historical kind, that makes it elusive.

Many political disagreements are, at bottom, disagreement about facts. In this context, one might argue that even though "truth" might remain practically out of reach, at least it makes sense to look for it (e.g., to answer the question "Has the euro made life more expensive in the countries where it has been introduced?"). Disagreements between political opponents should ideally be resolved when enough accurate information is gathered, and when people make an effort to overcome those cognitive biases that make them blind to the available evidence. This means

---

[32] One major problem with causal claims is that they are always "all things being equal otherwise"—and even if people agreed on social scientists' verdict, they could always argue about whether the latter condition obtains.

that an epistemic argument for democracy at least makes sense for factual questions, proposing that democracy is good, and possibly better than alternative regimes, at processing information about facts and reaching an overall correct or true picture of reality as it can be known.

Even if the epistemic argument for democracy suggested only this much, it would already be a major achievement. Disagreements over facts are one of the main reasons why partisans cannot agree on policy issues. In many cases, if people could only agree on the facts, they might discover that they do not disagree on much else. So, if involving the many were a way to get at this kind of truth, it would already provide a good reason to support democratic decision making.

3.2 Two Types of Values

Many political questions, however, are not purely factual and also involve normative principles or values. The standard of correctness for values is much trickier to conceptualize than it is for facts. Here I need to introduce a distinction between fact-sensitive or context-dependent political principles and basic values, a distinction I borrow from G. A. Cohen (2003) and modify for my purposes. I will argue that the standard of correctness for context-dependent political principles is a combination of empirical accuracy and coherence with underlying basic values. Whether one also posits a standard of correctness at the level of basic values determines the difference between culturalist and absolute cognitivism.

3.2.1 CONTEXT-DEPENDENT POLITICAL PRINCIPLES

Borrowing from G. A. Cohen's (2003) category of "fact-sensitive principles," I define "fact-sensitive" or, more generally, "context-dependent" political principles as political principles whose normative value can be at least partially justified by certain facts when those are verified. Thus, *if* it is the case that Iran's nuclear program threatens American security and world peace in general, then this fact is a reason to support the political principle that the United States ought to prevent Iran from acquiring the nuclear bomb. Similarly, *if* it is the case that a certain level of minimal wage tends to cause more unemployment, in a context of high unemployment, then this causal relation is an argument, if not against the principle of the minimum wage itself, at least against raising it above a certain threshold. Finally, *if* legal prohibition demonstrably encourages marijuana consumption, then this fact is a reason, all things being equal otherwise, to legalize it. Notice that while taking these "facts" into account gives reasons to endorse a given principle, they are not the only reasons, nor are these reasons necessarily sufficient. Nonetheless, facts can form part of the justificatory and explanatory basis of a given principle.

Again, here, to the extent that people will agree on the facts, it is likely that they will tend to agree on the fact-sensitive principles as well. Fact-sensitive principles, one might say, are partially correct to the extent that they have a true empirical basis. Conversely, some people disagree on fact-sensitive and context-dependent principles because they disagree on facts. A lot of apparent "value pluralism"—for example, the disagreement between Democrats and Republicans over the legitimate size of government—can arguably be explained by a disagreement about facts, including complicated facts such as causal relationship between big government and efficient spending in a given social and economic context. This disagreement about the facts of the world leads to a disagreement about political principles that are dependent on these facts. At the bottom, however, both Democrats and Republicans may share a common core of basic normative principles.

3.2.2 BASIC VALUES
Some of the context-dependent political principles can arguably themselves be traced back to a more fundamental set of normative political principles, in the same way that in Cohen's typology, fact-sensitive principles can be traced back to "basic value judgments." According to Cohen, basic principles account for why a fact can be treated as a reason to support the lower-order, fact-sensitive principle, without this basic principle being itself supported by either a fact or a more primordial principle. To go back to our previous examples, the higher-order principle to which one would ultimately regress in order to justify taking actions against Iran's nuclear program would be something like the principle that "Absent other considerations, human life should be preserved." Intermediate facts and principles may intervene—such as the likelihood of success of the US efforts to stop Iran from developing the bomb, or the intermediate principle that "National security trumps respect for another nations' right to noninterference"—but ultimately some kind of original grounding must be found, which, according to Cohen, is a fact-insensitive principle. Unlike the fact-sensitive principle that advocates limits on nuclear proliferation, which is contingent on the danger of nuclear weapons and perhaps the effectiveness of putting limits on their number, the principle that "Human life should be preserved, absent other considerations" is not rooted in any further "fact."[33] The parallel with G. A. Cohen's basic

---

[33] Even if an objector could point to a fact in which the value of human life was rooted, Cohen would argue that this fact itself would be relevant only when connected to a fundamental, fact-insensitive principle, beyond which no more regression would be possible. Against the possibility of infinite regressions, Cohen advances three arguments. First, it is just implausible that a credible interrogation of that form might go on indefinitely. Second, it is implausible that there exists the infinite nesting of principles that an infinite regression

value judgments ends here, however, because I think it safer to remain
agnostic as to whether the foundational norms of our political systems—
constitutional principles—are fact insensitive in the way Cohen argues
that basic value judgments are.

4. Political Cognitivism: Culturalist versus Absolutist

Here I need to complicate the picture somewhat by adding to the distinction between weak and strong political cognitivism an independent
distinction specifying the relative or absolute value of the standard. A
procedure-independent standard of correctness is by definition independent of the decision procedure. It can be, however, dependent on other
things. Since public values provide the standard for the value part of
political decisions, something needs to be said about the status of these
public values.

A political cognitivist has two choices at this stage. She can take for
granted a given set of values shared by a community and take this as the
touchstone of "correct" political judgments. In that case, the procedure-
independent standard is not independent of anything else but rather
determined by a given context, history, and culture. On that view—the
"culturalist" view of political cognitivism—the standard is a socially and
culturally determined standard, an ideal image of what we think we are
or ought to be as a group. Rorty (e.g., 1979 and 2006), Walzer (1981),
and the later Rawls (1993) are arguably representatives of such relative
political cognitivism—with considerable nuances, of course. Rorty has
been characterized as an "Archimedean relativist" (Dworkin 1996: 88–
89), denying that there is any privileged moral point of view in the world.
Walzer simply acknowledges the value of our Western cultural heritage.
The later Rawls, finally, takes this heritage as a springboard for a "reflective equilibrium" and a constructed notion of justice, which may encom-
pass some less culturally relative notions. Nonetheless the later Rawls
reintroduced more relativistic elements by specifying that his theory of
justice was "just for us" (Western liberals), not just in the abstract or for
all human beings.

Alternatively, the political cognitivist can posit the existence of a standard of correctness at the meta-level of basic values. In other words, at
this stage, political cognitivism is premised on moral cognitivism, the idea

---

would require. Third, the justificatory sequence cannot proceed without end because our
resources of conviction are finite, and even if they were not, proceeding without end would
violate the self-understanding stipulation (one must have a clear grasp of the reasons why
one holds a principle) (Cohen 2003: 218).

that basic normative principles themselves can be assessed from the point of view of an independent standard of moral correctness that is universally valid. This is what might be called the "absolutist" view of political cognitivism. Habermas and the early Rawls may arguably count as representatives of such absolutist political cognitivism (at least, in their case, at the level of constitutional norms and to the extent that their ideal procedures—deliberation in the ideal speech situation or in the original position—amount to a standard of moral truth). The paradigmatic case for absolutist political cognitivism, however, is Plato.

Culturalist political cognitivism consists of positing that the standards of correctness for any political decisions are public values. Political decisions are correct when they not only rest on accurate empirical information but also match and cohere with the set of existing public values, as they are defined at a given place and time. The correct answer is not correct in the abstract but correct "for us." In that view, the standard is a socially and culturally constructed standard, an ideal image of what we think we ought to be as a group. Culturalist political cognitivism is compatible with almost any metaethical positions, including moral noncognitivism. Wherever the basic set of public values comes from (history, moral "facts," intuitions, emotions, reason, or intersubjective agreement), what matters is that one may speak of a correct answer at the level of the group.

Absolutist political cognitivism, by contrast, posits the independent standard of correctness as something distinct from any given public values. Absolutist political cognitivism is premised on moral cognitivism. In that view, policy decisions are correct or right not only to the extent that they rest on accurate empirical information but to the extent that they cohere with a set of transcendent moral norms.

On a weak interpretation of those ends (avoidance of errors), absolutist political cognitivism would most likely define the standard as avoidance of famines, wars, genocides, and other catastrophes. A political decision could thus be said to be correct whenever it satisfies, or is compatible with, avoidance of such harms (see also Estlund 2008: 155).[34] On a strong interpretation of the standard, a possible interpretation could be defined as, say, Pareto optimality: a political decision would be correct whenever it improved at least one person's welfare without harming anyone else's. The criterion of Pareto optimality, however, suffers from the difficulty that in practice, almost no political decision has the property of enhancing everyone's welfare at no cost to anyone else. There is always at least someone, and generally a group of people, who is harmed by a change in the status quo. The criterion of Pareto optimality sets the

---

[34] Estlund offers a list of primary calamities as "war, famine, economic collapse, political collapse, epidemic, genocide."

highest possible bar for the possibility to call a decision correct. Nevertheless, one can see it as a regulative, ideal upper bound for the domain of correct political decisions. A perhaps less exacting standard might be the standard of efficiency as defined by classical utilitarianism, in which a decision leading to some gain for society as a whole can be good even if it harms some people in the process.

5. Implications for the Epistemic Argument for Democracy

The epistemic argument for democracy will vary in strength depending on the kind of political cognitivism it is premised on. For weak political cognitivists, the epistemic argument for democracy simply asserts that democratic decision making is good at avoiding bad outcomes, whether defined in relative or absolute terms. For strong political cognitivists, the epistemic argument asserts that democratic decision making is good at producing positive answers to given problems, where, again, these answers may be culturally relative or universally valid.

My own position on this could be characterized as a hybrid form of strong political cognitivism, which includes culturalist views for some questions and absolutist views on others. I believe that democracy is good at producing the right kind of outcomes, where the "right kind of outcomes" is defined in part in relation to a shared set of public values that cannot claim universal but merely local validity (e.g., a certain view of the hierarchy between equality and freedom) but in part, also, with a smaller core of values that have universal validity (e.g., the ideal of human rights and a number of basic freedoms). This hybrid version of political cognitivism seems to me a more compelling representation of how we actually think of our collective values.

6. Status of the Standard: Postulate or Empirical Benchmark?

Another question that can be raised about the independent standard of correctness is whether it is merely an assumption that we make in politics in order to make sense of our existing practices—through a transcendental deduction of sorts (going back to the condition of possibility of a given practice)—or whether it is an identifiable, empirically measurable standard. For the sake of the epistemic argument, it only matters that it be a plausible assumption. The independent standard of correctness is something we postulate every time we debate and vote in the hope of finding a solution to a problem. Yet it is not impossible, and would in fact prove very useful, to think of it as something that can be quantifiable.

As an assumption, the independent standard of correctness is widely accepted outside of democratic politics: from Plato on, the elitist tradition that promotes kingship or aristocracy, and that today promotes something that might be called "expertocracy," or the rule of experts, has always assumed that politics was a matter of knowledge and expertise—that one could therefore differentiate between good and bad political judgments and, consequently, right and wrong decisions. In the context of contemporary deliberative theories, as we saw in chapter 2, this acknowledgment—that one can somehow tell the difference between good and bad judgments—is buried under concerns for the fairness of democratic procedures, as if it were politically dubious to raise the question of the people's competence or the efficiency of democratic procedures at reaching good outcomes. Even there, however, one can demonstrate that the independent standard of correctness is postulated as an assumption needed to make sense of deliberation and even voting (see Marti 2006; Estlund 2008). The independent standard of correctness is thus an (unacknowledged) mainstream assumption in politics and an implicit element of even theories of democracy that appear purely procedural. In summary, the independent standard of correctness is the condition of possibility of any normative thinking about politics. Being a postulate, however, the independent standard of correctness need not be an empirically identifiable standard.

There has been, however, at least one interesting attempt at quantifying the standard of good political judgment, which illustrates the tangibility of the idea of an independent standard of correctness. In his book *Expert Political Judgment* (2005), the political psychologist Philip Tetlock attempts to quantify the goodness, that is, the accuracy or rightness, of political experts' judgments.

Wrestling with the complexities of the process of setting standards for judging judgment, Tetlock mentions that the main difficulty lies in the relativist and skeptical opponents' claim that there is hardly "a topic more hopelessly subjective and less suitable for scientific analysis" than political judgment (p. 3). As Tetlock concedes, it is not easy to make room for some convergence on the characteristics of good political judgments, while at the same time acknowledging that disagreement about whether a particular judgment was actually good or bad, right or wrong, is irreducible.

Tetlock attributes this difficulty to the fact that political judgments, more so than any other form of judgments (such as legal or medical judgments, which political judgments are usually unfavorably compared with), hinge on hard-to-refute counterfactual claims about what would have happened under a different policy path and on impossible-to-refute moral claims about the types of people we should aspire to be—in other words, all claims that partisans can use to fortify their positions against falsification (p. 4). Despite these difficulties, Tetlock argues that we ought to resist the temptation of moving toward full-blown relativism, and struggle to identify some uncontroversial criteria of good judgment.

The two elements that Tetlock finally settles upon are empirical accuracy and logical rigor. The standard of empirical accuracy requires that the private beliefs of political experts map onto the publicly observable world and satisfy correspondence tests rooted in empiricism. The standard of logical rigor requires that political experts have internally consistent beliefs satisfying coherence and process tests rooted in logic. Experts should have coherent systems of beliefs and update these beliefs according to available evidence. In other words, good political judges should both "get it right" and "think the right way" (p. 7).

The way Tetlock proceeds in order to assess experts' political judgment is by testing the factual forecasts on which these judgments are premised against the actuality of what has happened. Tetlock (2005: chaps. 2 and 3) considers the past history of dozens of countries and topics such as transition to democracy and capitalism, economic growth, interstate violence, and nuclear proliferation. The implicit idea is that when an expert recommends a given course of action, it is in part because of the way he or she thinks the world will turn out to be. In other words, the expert's principled judgment (which is political to the extent that it has this principled, normative dimension) hinges at least in part on a factual dimension. Because of that explicit linkage between the normative claim and the factual forecast, Tetlock can retrospectively assess the validity of the policy recommendation, after the predicted course of event has taken place or failed to do so.

Consider the following examples. It is a fact that Hitler posed a threat in the early 1930s. In light of that evidence, the appeasement policy advocated by many politicians within European democracies at the time can be said to have been shortsighted and, in effect, mistaken and wrong. Similarly, because we now know that the Soviet Union was vulnerable in the early 1980s, it can be argued that Reagan rightly played the deterrence card when Cold War doves wrongly built up the risks of a conflict. More recently, after 9/11 revealed to the world how high the terrorist capabilities of radical Islamic organizations in the 1990s were, both Republican and Democrat administrations can be at least partially blamed for a certain amount of negligence in protecting national security. Conversely, those people who were capable of foreseeing these events long before the rest of us and made political recommendations accordingly can be credited with good political judgment.

Of course, these judgments are controversial. People will, in effect, even argue on the factual dimension of the judgment. For example, ex-doves still argue that the USSR was a formidable threat during the Cold War and that just because we avoided a nuclear war does not mean that a rerun of history would not yield a catastrophic outcome. It is not that the risks were not high; it is just that we got lucky. Historical facts are always subject to endless interpretation and the question of their truth never fully settled—a possibility that undergirds Hannah Arendt's already mentioned point that factual truth is even more vulnerable and liable to distortion than the rational truth of mathematic axioms, for example. If historical facts are endlessly controversial, policy recommendations based on a contemporary assessment of those facts are bound to be even more so.

To counter this type of objection, Tetlock resorts to what he calls a "multi-method triangulation"—a complex method mixing quantitative and qualitative techniques—to refine his definition and his standard of a good political judgment. One of the most interesting aspects of this method is that it both makes room for the point of view of the subjects themselves and introduces some flexibility in the accuracy diagnostic. Tetlock, for example, factors into the equation variables such as *value adjustments* that respond to forecasters' protests that their mistakes were the "right" mistakes, given the cost of erring in the other direction; *controversy adjustments* that respond to forecasters' protests that they were really right and Tetlock's reality checks wrong; *difficulty adjustments* that respond to protests that some forecasters had been dealt tougher tasks than others; and even *fuzzy-set adjustments* that give forecasters partial credit whenever they claimed that things that did not happen either almost happened or might yet happen.[35]

In the end, Tetlock derives convincing conclusions about (the limits of) political expertise. One of the most interesting results—as discussed in brief above—shows that of the two types of political experts identified by Tetlock (and labeled after Isaiah Berlin's famous ideal types), "foxes" almost always outperform "hedgehogs." Foxes' knowledge of many little things, their ability to draw from an eclectic array of traditions and to improvise in response to changing events make them better forecasters than hedgehogs, who know only one big thing, toil devotedly within one tradition, and impose formulaic solutions on ill-defined problems (Tetlock 2005).

By defining his standard of good political judgment as a combination of empirical accuracy and formal rigor, however, Tetlock avoids the question of the rightness of the basic values supporting a political judgment. The quality of policy recommendations is judged not on the merit of their underlying basic values but on whether or not the facts and events they hinged upon failed to take place and, thus, whether or not people had coherent systems of beliefs within a given frame of values. For example,

---

[35] For more, see Tetlock 2005, technical appendix.

Cold War doves are retrospectively judged as wrong to the extent that they overestimated the risks of a war with the former USSR, but not because of their intrinsic views about the value of cooperation and pacifism. Similarly, hawks are judged to be right to the extent that they correctly assessed the military superiority of the Western camp, not because of their moral views about deterrence and massive retaliation.

Tetlock thus stays clear of moral cognitivism, to the point where he cannot even criticize the foundations of a number of policies generally acknowledged as morally wrong. If Hitler's political judgment were to be criticized, on Tetlock's terms, it would be for its lack of empirical grounding (e.g., the lack of biological evidence behind the ideology of Aryan supremacy) or its failure at weighing beliefs correctly (illustrated by the miscalculation of a US involvement in the case of an attack on its European allies). Hitler's political judgment, however, would not be faulted for its ideological and moral foundations, such as racism, eugenism, or warmongering. Although he does not want to settle for full-blown relativism, Tetlock remains de facto agnostic in the sphere of what he calls "morals."

In restricting himself this way, Tetlock keeps his claims minimally controversial. There is a sense, however, in which he is only able to come up with a limited way of assessing political judgment. If we allow some room for basic values—for example, along the lines of the hybrid form of political cognitivism that I favor myself—even at the risk of introducing more indeterminacy and controversy, then at least we have a fuller account of the "rightness" or "wrongness" of political judgment.

7. The Antiauthoritarian Objection

While quantitative political scientists may be constrained by practical considerations to stay away from the moral dimension of political decisions, normative theorists may be tempted by an explicitly moral approach. Culturalist political cognitivists will define the standard as relative to a given people. Absolutist political cognitivists will even more ambitiously define the standard of correctness as universal. The absolutist interpretation of the independent standard of correctness may seem to equate democratic decision making with a quest for moral truth, which might raise some concerns. These concerns can be found in particular in the work of two very different authors, Hannah Arendt and John Rawls, who nonetheless share a distrust of the concept of truth.

Truth has, it must be said, a bad reputation in political theory (much more so, in any case, than in philosophy). Even barring the implausible position according to which politics is simply about the fair adjudication of incommensurable preferences—and applying the vocabulary of truth to preference aggregation is, consequently, a categorical mistake—there are two prominent groups of skeptics in political theory. On the one hand, one finds postmodernist relativists, who may grant that there is something like a common good for a given community but would deny that the concept of truth aptly characterizes it. Although relativists' epistemological quarrel with truth is interesting, I will not enter that debate, as this would take me too far afield and probably would not settle the issue anyway. On the other hand, one finds the category, more relevant for my purposes, of the "liberals," who may or may not share relativists' skepticism about the relevance of truth for politics but instead mostly worry (like many postmodernists) about the potentially authoritarian implications of that concept. Because truth is invoked to silence dissenters as "mistaken" and to artificially pacify the messy reality of a free society, it is generally seen as the Trojan horse of totalitarianism. Liberals thus share with some postmodernists the belief that the vocabulary and the concept of truth must be kept out of politics because whether or not truth turns out to be relevant as an epistemological category for the study of politics, it is at any rate too dangerous. I now consider two authors particularly worried by the illiberal implications of truth: Hannah Arendt, who in some way foreshadows contemporary agonistic pluralists (a certain type of postmodernists), and John Rawls, who represents the camp of liberal pluralists. Truth, both Arendt and Rawls argue, has a coercive nature inimical to the spirit of tolerance and accommodation necessary to political life. In what follows, I attempt to demonstrate that this worry is exaggerated and that there should be room, in our political discourse, for a concept of truth amicable to the spirit of tolerance and healthy skepticism ideally part of a democratic ethos.

7.1 Arendt on Truth and Politics

In her essay "Truth and Politics" (in Arendt 1993: chap. 7), Hannah Arendt distinguishes between two kinds of truth: factual and rational. Factual truth includes truth about facts and events. Rational truth includes mathematical, scientific, and philosophical truths. This distinction, which Arendt takes up wholesale from the philosophical tradition, allows Arendt to diagnose two types of tension in politics: a fundamental tension between totalitarianism and factual truth on the one hand, and a tension between politics and rational truth on the other. Arendt thus writes against two different dangers. The first danger is the denial of factual truth, which happens either in a totalitarian context, when the government manipulates information and rewrites history, or in a democratic context, when knowledge is trivialized into "opinion" (so that, for example, a denial of the Holocaust is defended as "a right to one's opinion"). The second type of danger Arendt worries about is the suppression of debate that occurs when claims to rational truth monopolize the terrain of politics and morality.

Arendt's concerns lead to a double, implicit assertion: a sound politics should be concerned about the truth of facts, particularly historical facts—a truth that is fragile and in need of protection. Thus, for Arendt, "freedom of opinion is a farce unless factual information is guaranteed and the facts themselves are not in dispute" (Arendt 1993: 232). On the other hand, a sound politics should not get involved in debates about rational truths, particularly not those of the moral type (if they exist—a possibility that Arendt neither asserts nor rules out), because truth claims in that realm are coercive by nature and leave no room for debate. In politics, Arendt suggests, opinion, not truth, should reign free.

While Arendt might not be worried about an epistemic argument for democracy that would simply apply to the factual side of political questions, she would probably vehemently oppose the idea that the many, or anyone for that matter, can claim any knowledge of what is morally and/ or politically "right." Arendt would thus refuse political cognitivism, not because of a relativist conviction that there are no such things as right and wrong answers in politics, but because the endorsement of political cognitivism would undermine the very essence of politics. If politics is about debate and disagreement, truth claims can only put an end to it.

This position is, in many ways, that of a number of agonistic pluralists (e.g., Sanders 1997; Mouffe 1998), whose views also strive to preserve the political space as an arena for open disagreement and endless debate. Arendt's argument thus stands as a powerful objection to deliberative epistemic democracy, among other things. There are, however, also limits to a position that peremptorily denies the possibility of a nonauthoritarian quest for truth. In effect, the epistemic argument for democracy does assert that the rule of the many is, in part, such a nonauthoritarian, collective quest for truth.

The first thing one might criticize in Arendt's position is the dichotomy of factual and rational truth. As we saw earlier, factual truth can be fairly complicated. When it comes to complex causal relationships in particular, the difference between this type of "fact" and rational truths of the scientific kind is simply a difference between genus and species, a scientific law and its application. So if politics should be concerned about factual accuracy, it should also be concerned with the rational conditions for that accuracy. If that is the case, then politics cannot afford to be as separate from sciences and the conditions of production of knowledge as Arendt would like.

One may also be concerned with the way Arendt lumps together mathematical, scientific, and philosophical truths as "rational," and summarily contrasts them with "opinion." This Platonic contrast, which Arendt criti-cizes and yet reproduces in her own analysis, has been challenged in sci-ences and philosophy at least since the probabilization of knowledge in the mid-eighteenth century. Few people today would argue that the truths of science and philosophy are as absolute as mathematical tautologies. In the scientific, and even more so in the moral realm, we are, at best, the bearers of probabilistic, fallible truth. One might thus argue that the difference between truth and opinion in politics is merely one of degree, such that philosophical and rational truths do enter politics, even if only in the guise of opinions. This also has implications in terms of the division of labor between "experts" and "amateurs." If there is only a difference in degree between truth and opinion, then the question of who has authority to decide what depends on the probabilities of knowing the right answer assigned, respectively, to the elites of knowledge and mere citizens.

Third, once the Platonic distinction between truth and opinion is no longer taken for granted, one may question Arendt's assertion about the "coercive nature of truth." Arendt wrote in the lingering shadows of totalitarian politics. Because of this historical vantage point, she may have come to see an essential connection between totalitarianism and truth claims, where there might simply be a contingent one. Two genera-tions before Arendt, for example, as we saw in chapter 3, John Dewey saw an essential connection between the quest for a pragmatic form of truth and . . . democracy.

I would thus argue, contra Arendt, that there is a nonauthoritar-ian way to make truth claims. Opinions, however mistaken they may ultimately be, are tentative truths, not just arbitrary manifestations of the will. When expressed as such, opinions correspond to, and shape, a certain type of political behavior that might be called "democratic." It consists in being respectful of other people's opinions and skeptical of one's own, while stopping before the edge of full-blown relativism. When entering political debate, we throw our opinion into the discussion to see whether it resonates like truth when in contact with other opinions and arguments. Aware of the fact that the margin of uncertainty is important, however, we are, or should be, willing to listen to the other side and occa-sionally yield to it.

The difference between democracy and totalitarianism might thus not be, as Arendt thought, that democracy is not concerned with ratio-nal truth, but that it has a different attitude toward it. When Habermas evokes, in the moral sphere, "the unforced force of the better argument," the kind of coercion expected is the compelling and nonviolent force of the "right," which, for Habermas, is analogous to truth. Speaking as if there was a right answer is, in that case, a promise of good politics. An effort to get at the truth, or some kind of independent standard of correctness, need not be coercive. The point of democracy might be that when we disagree with someone, we make arguments that we think in good faith must be right and ought to convince anyone who is exposed to them. But we ought not be so self-assured as to be ready to kill everyone who disagrees, or die rather than change our minds, and generally refuse discussion and a chance to listen to the opposite views. In that sense, truth claims in a democratic context need not exclude the possibility of debate. In effect, if one of the reasons why we endorse democracy is because of what we believe to be the power of "democratic reason," then we should consider ourselves as pieces in a larger picture and make assertions based on the assumption that our views are only partially right.

Arendt's conviction that there is something coercive about truth is, in effect, a half-truth, and one that can be questioned. Truth is never coercive by itself, not even mathematical truth. Only dictators are. Thus, we should not necessarily give up on truth in politics altogether. Democratic politics in particular might keep truth as an ideal, as long as it is a probabilistic, modest, and, most importantly, unarmed kind of truth.

7.2 Rawls's Epistemic Abstinence

I have just offered an answer to the fear that the idea of an independent standard of correctness inspires in politics by denying that even if we call it truth, it has the coercive nature someone like Arendt attributes to it. Another strategy, however, is to avoid the word "truth" altogether and substitute for it another concept that performs the same function of an independent standard of correctness without, however, bearing any of the absolutist connotation conveyed by the concept of truth. Rawls's constructivist notion of the reasonable arguably offers such an alternative.

The later Rawls shares with Arendt a concern that truth claims in politics breed intolerance, making it impossible for people of different confessions and worldviews to live in peace with each other. To the extent that disagreement about values cannot remain private but must be managed at the collective level in the common space of public reason, Rawls proposes to apply there what he calls epistemic abstinence, substituting the notion of the reasonable for the concept of truth. Epistemic abstinence consists in refraining from claiming that one's doctrine of justice is true. This epistemic abstinence actually represents a subtle shift from Rawls's earlier claims. In A Theory of Justice (1971), if Rawls did not explicitly present the two principles as universally true, he nevertheless offered them as the foundation of any ideal just society. In fact, A Theory of Justice was fraught with truth-like moral claims. For example, the Rawls of a Theory of Justice was not in the least shocked by Condorcet's assumption that one could be correct about moral and political decisions, and he seemed to take for granted that there were "right" answers to such questions. His only worry regarded the implausibility of the independence assumption (Rawls 1999: 314–15).

In the later Rawls's view, and that of some commentators, epistemic abstinence represents a necessary acknowledgment of the fact of value pluralism and the constraints this fact puts on political theory. Under circumstances of deep and irreducible disagreement about ends and the meaning of the good life, there is just no room for truth claims about justice.[36] This applies of course to Rawls's theory itself. Contrary to what was perhaps suggested in *A Theory of Justice*, the later Rawls does not think that his theory of justice is the right one, let alone the true one, but merely the best for us—where "us" is defined as the members of modern, constitutional democracies like the United States. *Political Liberalism* (1993) thus remains agnostic as to the truth value of any of its political assertions.

Rawls's theoretical reason to abjure truth is that claiming truth for a political conception of justice or claiming that its truth is reason for accepting it would be self-defeating from the point of view of a theory of justice that precisely tries to transcend disagreement about the right and the wrong, the true and the false, and any metaphysical doctrines. In order to be valid as a theory of justice, Rawls's theory of justice must renounce truth, replacing it with a metaphysically lighter notion: the reasonable. Because of that substitution, Rawls can argue that his conception of "political, not metaphysical" liberalism does entirely without the concept of truth. Public reason is not public truth, nor is it, for that matter, its opposite. Public reason has nothing to do with truth.[37]

Not all commentators have greeted the move toward epistemic abstinence as progress. This move indeed raises some important questions about the conceptual relationship between truth and the reasonable (see Raz 1990; Habermas 1995; Estlund 1998; Cohen 2009). Habermas is probably here the most helpful commentator, offering two possible interpretations of this conceptual relationship. According to Habermas: "Either we understand 'reasonable' in the sense of practical reason as synonymous with 'morally true' that is, as a validity concept analogous to truth and on the same plane as propositional truth. . . . Or we understand 'reasonable' in more or less the same sense as 'thoughtfulness' in dealing with debatable views whose truth is for the present undecided"

---

[36] In fact, for the Rawls of *Political Liberalism*, truth is now an apple of discord on a par with comprehensive worldviews and thick notions of right and wrong. Truth bars the possibility of a common ground among people of different traditions of thought or religious backgrounds.

[37] Nevertheless, regarding at least moral basics, even the Rawls of *Political Liberalism* seems to acknowledge that an overlapping consensus might be indicative of something like the "truth."

(Habermas 1995: 123–24). The first option is the one favored by Habermas, and indeed it is the one that most resembles his own position on the status of moral claims versus truth claims. The reasonable is analogous to the true, but it applies to different objects and it describes a different type of relation between the concepts and reality.

The second option, however, is that favored by Rawls, and one that Habermas has difficulties making sense of. The reasonable for Rawls would be some kind of provisional status for views that have not yet been proven wrong and may or may not be proven true in an indeterminate future. Further, reasonable views are characterized by the type of people holding them—to wit, people who are willing to put up with the undecidedness of such views. The reasonable is thus "a higher-level predicate concerned more with 'reasonable disagreements,' and hence with the fallibilistic consciousness and civil demeanor of persons, than with the validity of their assertions" (Habermas 1995: 124).

This fallibilistic dimension of the reasonable applies in particular to claims about the demands of justice. What is interesting, as Habermas remarks, is that "a reasonable conception of justice preserves an oblique relation to a truth claim projected into the future" (Habermas 1995: 125). It is because access to this future truth is forever postponed in a world characterized by deep disagreement about values that we have to fall back on tolerance toward worldviews that are not themselves patently unreasonable. When push comes to shove, then, it seems that "not patently untrue" forms a possible content of the notion of reasonable.

Rawls's answer to Habermas confirms the accuracy of the latter interpretation. Rawls repeats that the reasonable, unlike the true, is a concept that recognizes the burdens of judgment and fosters liberty of conscience and freedom of thought. Thus, "the reasonable does, of course, express a reflective attitude to toleration" (Rawls 1995: 150, referring to Rawls 1993: 54–61). The reasonable is not the equivalent of a truth-like claim in the moral sphere. Further, Rawls insists on the specificity of reasonableness as a standard of objective validity without epistemic meaning. Even though Habermas (and others such as Raz or Cohen) think that political liberalism cannot avoid the question of truth, Rawls's answer is deceptively simple: "I do not see why not" (Rawls 1995: 150). The only point that Rawls seems willing to concede is that "the idea of the reasonable needs a more thorough examination than *Political Liberalism* offers." He also suspects that "people will continue to raise questions of truth and the philosophical idea of the person and to tax political liberalism with not discussing them." However, he concludes, "in the absence of particulars, these complaints fall short of objections" (ibid.).

This discussion of the conceptual relationship between truth and the reasonable in Rawls helps us figure out whether political cognitivism is compatible with the fact of value pluralism. First, if Rawls is right, acknowledging the fact of value pluralism is not incompatible with positing some kind of independent standard of correctness, such as the reasonable. Political decisions are just to the extent that they do not violate this standard of reasonableness. This may sound like a very limited criterion, compared especially with thicker conceptions of the independent standard of correctness, such as Pareto efficiency, but it is one nonetheless.

Second, this discussion establishes that abstaining from using the concept of truth is distinct from abstaining from positing the existence of an independent standard of correctness. Rawls's epistemic abstinence is, consequently, compatible with a form of political cognitivism (one that operates at the level of constitutional norms, however, not daily politics). Rawls's moral constructivism ensures this compatibility, as long as the concept of the reasonable is used in lieu of the concept of truth.

Whether one ultimately calls the independent standard of correctness truth, rightness, impartiality, reason, or even rationality,[38] might be a simple matter of semantics. My own view is that the concept of truth has a discrete and binary quality that is ill suited to describe the nature of human actions and choices, which more often fall short of perfection without being completely wrong. One may thus agree with both Habermas and Rawls in staying away from the vocabulary of truth, and opting for that of the "right" or the "reasonable" instead. I prefer to speak of "smart" or "intelligent" decisions myself.

One need not buy into a political conception of truth to accept the idea of a procedure-independent standard of correctness by which to assess the outcome of a decision procedure. In fact, one need not actually define what the nature of this independent standard of correctness might be. As I hope to have shown, there exist many theories of what this independent standard of correctness might be (whether it is a procedure, an outcome, a range of outcomes, a negative definition of what to avoid, etc.), and the reader should feel free to adopt any single one of them. The point here is simply to suggest that the burden of the proof is on those who deny the meaningfulness and the possibility of the existence of such an independent standard of correctness.

CONCLUSION

The goal of this chapter was to render the assumption of a standard of correctness independent from the decision procedure palatable to

---

[38] All those terms are used in the literature on deliberative and epistemic democracy, making for a lot of confusion.

skeptical readers, by insisting that political cognitivism need not entail a number of difficult positions, such as moral realism, or political authoritarianism. As we saw, political cognitivism can be weak or strong, culturalist or absolute. Only the absolute version excludes some metaethical views, such as pure value subjectivism and moral skepticism. On any other reading, political cognitivism is compatible with all sorts of metaethical views. I have also shown how, while the procedure-independent standard of correctness is primarily meant as a postulate, there are ways to quantify it (at least on the culturalist reading of political cognitivism) that potentially render the epistemic claim about the superiority of democratic procedures over nondemocratic ones empirically falsifiable.

Conclusion: Democracy as a Gamble Worth Taking

"The reason why our political system was superior to those of all other countries was this: Our state . . . is not due to the personal creation of one man, but of very many; it has not been founded during the lifetime of any particular individual, but through a series of centuries and generations."

—Cato

"The way Washington works is you often start with what's optimal, a best solution to some complex problem, and, surprisingly, there's often quite a bit of bipartisan consensus on what will actually work, at least in private. That's your herd of gazelles. But you've got to get them across the savanna safely, to a distant watering hole. And the longer it takes, the more you lose. You may end up with very few. You may lose them all. Because there are predators out there, lions and tigers, packs of hyenas, and they're big and fast and relentless—considering how any significant solution to a big problem is bound to be opposed, do or die, by some industries or interests who'd figured out a way to profit from the ways things are, even if they're profoundly busted! So that's your challenge: see how many gazelles you can get to the watering hole."

—Gary Gensler[39]

1. Summary

This book has proposed a sustained epistemic case for democracy. I have argued that there are good theoretical reasons to believe that when it comes to epistemic reliability, under some reasonable assumptions, the rule of the many is likely to outperform any version of the rule of the few, at least if we assume that politics is akin to a complex and long enough maze, the knowledge of which cannot reside with any individual in particular or even just a few of them. When the maze is complicated and long enough, the likelihood that the group makes the right series of choices that will ultimately get them out of the maze is higher when the decision is made in an inclusive fashion, pooling everyone's information,

---

[39] Cited in Suskind 2011: 395.

arguments, and perspectives, than when it is made by one member of the group only or just a few of them.

In order to make that claim, I have resorted to the idea of the emergent phenomenon of collective intelligence and, more specifically, to a concept of collective distributed intelligence of the people in politics, or what I have termed "democratic reason." This strategy allowed me to circumscribe the problem of the well-documented ignorance of the average citizen by arguing that democracy can function even when the individual citizens' input is of limited epistemic quality. I have argued that a certain number of democratic mechanisms turn citizens' judgment into collective decisions of higher epistemic value. The main two mechanisms that this book has considered are inclusive deliberation (among the people, where direct democracy is feasible, or their representatives for most other cases) and simple majority rule.

I have also argued that democratic reason can be properly channeled by deliberation and majority rule only to the extent that the inclusion of large numbers of people brings with it enough cognitive diversity. A large group is more likely to be cognitively diverse than a smaller subset of the same group, but the requirement of sufficient cognitive diversity also implies that the rule of the many occurs against the backdrop of a liberal society, characterized among other things by the existence of a Millian "market of ideas," independent media, a diverse economy and social life, and an education fostering the relevant democratic competencies. I now turn to these liberal preconditions for democratic reason.

2. Preconditions of Democratic Reason

I have defended the importance of cognitive diversity for the emergence of the phenomenon of collective intelligence. Whether it is even more important than individual abilities (in deliberative problem solving) or just as important (in judgment aggregation), cognitive diversity is thus crucial to democratic reason. Without it, the mechanisms of deliberation and majority rule risk producing democratic unreason. The precondition for the emergence of democratic reason is thus a context of respect and encouragement for cognitive differences. One is here reminded of Mill's argument about the epistemic value of diversity and the importance of institutionalizing a free market of ideas as key to the pursuit of collective truth (or at least the avoidance of errors). A free market of ideas ensures that the constant conflict of points of view and arguments renews perspectives, interpretations, heuristics, and predictive models—the toolbox of democratic reason. The emergence of democratic reason is thus conditional on the existence of a social and cultural context that nurtures and protects, among other differences, cognitive differences.

To the extent that the epistemic argument for democracy has value, therefore, it establishes that democracy and liberalism go together. In other words, democracy is likely to be smart only if it is also liberal. Illiberal or authoritarian democracies that foster conformism of views and stifle dissent risk turning both deliberation and majority rule into dangerous mechanisms for collective unreason, depriving themselves in particular of the possibility to come up with efficient solutions to collective problems, accurate information aggregation, and reliable predictions.

One key factor in that respect is the independence of the media, which plays a crucial role of counterpower but also interfaces between the public and government in representative democracies. Since the social aspect of freedom also has consequences for the type of education that is provided for future citizens, another implication of the claim about the importance of cognitive diversity for democratic reason is that the latter requires a type of liberal and democratic education as well—one that fosters creativity rather than obedience to authority, and teaches "democratic competences" such as the ability not only to express oneself but also to listen to others and respect divergences of opinion (see Dewey 1916, on the importance of a specifically democratic education; and Breton 2006, for a demonstration of the problems caused by the lack of such an education in contemporary democracies).

3. Limits of the Metaphor of the Maze

The metaphor of the maze and the masses introduced at the beginning of this book was meant to appeal to the reader's intuition about the advantages of group decisions over decisions made by one or just a few persons. When the maze is long and complicated, the likelihood that the group makes the right series of choices is higher when the decision is made democratically than when it is made by one or a few only.

Of course, the metaphor of the maze has its limits. I would like now to examine ways in which it is misleading, which will also provide an opportunity to review some of the main objections encountered in the book.

The first objection one might raise against the metaphor of the maze and the masses is in the form of a denial that politics is akin to a maze with an exit. Since we never get out of the maze, so the objector may say, it does not really matter which direction we take at every fork. This is partly true and partly wrong. It is true that solutions to collective problems are generally transitory—we never really get out of the maze. On the other hand, the example can be constructed in such a way as to fit more adequately with reality. Imagine that the group is now a tribe exploring a dangerous and complicated environment—humanity at its beginning in other words. There is no end to the exploration process. Still, on their nomadic and endless journey, the tribe faces choices that they can solve in a better or worse way. Let us say the tribe faces a choice between crossing a river and continuing to walk along it without crossing it. Depending on the width of the river, the temperature of the water, the time of year, the number of young children in the group, and so on, it might prove a fatal decision to cross the river. On the other hand, if crossing the river is not too dangerous and appetizing berries happen to grow on the other side, it could prove to be a better decision than staying put. One can argue that in the long run, the group as a whole is more likely to make the right decisions for itself than by letting one person in the group make all the choices.

To objectors who remain unconvinced that this kind of metaphor—the people as a group lost in a maze or a tribe exploring the savanna—applies to politics, or consider that it stems from a naively idealistic perspective of the reality of what politics is about, I will here only respond by referring them to a strikingly similar image used by Gary Gensler, a consummate political insider. Asked about his experience in Washington during Obama's first term, Gensler compares politics to getting a herd of gazelles to a water hole (see the second epigraph of this conclusion). While it may be the case that by the end of the political process, many gazelles are lost to lobbyists and other predators, and what is achieved by Congress has more to do with managing conflicts of power and interests, what we start with, initially, is the kind of collective problem solving that I have argued throughout this book is, and should be, central to democratic politics: getting the gazelles to the water hole or, in my own imagery, getting the group safely home, out of the maze, or at least to its next best destination.

A second objection argues that if we assume that there are several ways to exit the maze, then it must be the case that some choices are equally good and that the only value of democratic decision making in helping us choose among them is due to the intrinsic fairness of its procedures. Whether people should aim for exit A (on the left) or exit B (on the right) is a mere problem of coordination. Here, what democratic decision making has to offer is that it respects everyone's interest by including everyone's voice in the decision procedure. But we cannot say that democracy helps us make a "better" choice in any meaningful sense.

It is true that in some cases, the outcome of political decisions is neither "right" nor "wrong." This is what we assume when it comes to the choice of some fundamental values on which we must simply coordinate as a group. One might say that at least in some cases, for example, whether the government follows a right-wing policy or a left-wing policy is such a coordination problem. It is a choice between exit A and exit

B. We go with what the majority wants because that is the fairest way to determine the fate of the country, regardless of whether a right-wing platform is better or worse than a left-wing one. Other examples would be the level of risk that people are willing to accept (eating genetically modified organisms, learning to deal with nuclear waste, etc.), or what kind of trade-off between equality and freedom they are ready to strike. In the end the majority is generally what drives the choice for one value over the other (risk over safety, freedom over equality, or vice versa) as the fairest decision-making procedure.

Acknowledging that some choices are a matter of democratic fairness need not, however, lead us to deny that there are a large number of issues where we can speak of a better or worse answer. Once a country has settled for a fundamental value—say the principle of precaution over that of risk taking—a lot of disagreement about the best means to that end still needs to be resolved. One value can translate into many different policies. Or, in other words, once we settle for exit B, say, the question still remains: How do we get there? I have argued that even such apparently technical decisions benefit from popular input.

But second, even at the level of fundamental values, it is not so clear that all choices are equally valid. Without endorsing a form of moral realism, one might say that if the choice is between freedom of conscience and strict obedience to religious dogmas, between equality and a caste society, between human rights and slavery, there is a right answer and a wrong answer. So even though a certain ranking of freedom over equality might not be better or worse than another ranking of equality over freedom, notice that the two values at stake are freedom and equality, not slavery and racism.[40]

Thus, some trade-offs between the value of social justice and economic freedom may just amount to a simple coordination problem in which the benefits of democracy are purely procedural. Defining the set of values that we want to pursue as a society, however, is not just a matter of subjective preference. Picking slavery or a caste society is wrong. Even in the case of the death penalty or female genital mutilation, I would argue, there is a better and a worse position. As to the abortion debate, it is perhaps the heart of the matter, causing such passionate disagreement that liberal democracies have generally opted to take it out of the democratic

---

[40] In that sense, we are slightly blinded to the reality that some choices are worse than others when we compare only democracies, which for the most part (and in keeping with my thesis) have already made the "better" moral choices. Existing democracies vary in the ways they treat their "least well off," for example, but none of them has decided on an official policy of actively disenfranchising them.

debate. From a purely moral point of view, however, it is not clear that there is not a better and a worse answer there too.

To go back to easier cases, except for moral and cultural relativists perhaps, most people value freedom and human rights over slavery and genocide, and personal autonomy over blind respect for hierarchy and the law of the stronger. This goes to show that even in the sphere of values, preferences are not entirely subjective and random. It is thus plausible that on some fundamental issues at least, under conditions conducive to the full expression of cognitive diversity, democratic reasoning is more likely to yield decisions that are morally right than reasoning left to one or a few members of the group (even "moral" experts in some cases). To paraphrase Machiavelli, one might argue that the voice of the people is the voice of God, in the sense that when it comes to moral truth, we are more likely to find it in the people themselves than in the prince or oligarchs.

A third objection to the metaphor of the maze is that political choices are usually more complicated than a choice between two predefined alternatives. In politics, there may be an infinity of options to choose from, and not only do we have to choose from among them but we also have to identify and formulate them. So the metaphor of a fork in a maze is overly simplistic. That objection is correct. The point of a metaphor, however, is to offer a striking and concise expression of a more complicated thought. The simplicity of the metaphor does not make it entirely irrelevant or meaningless. In theory, any decision can be reduced to a successive choice between two options. Granted, this kind of subdivision then opens the decision process to the problem of agenda manipulation, which cannot be solved without introducing some amount of potentially epistemically costly indeterminacy. This is, however, a distinct problem from the theoretical problem just raised about the implausibility of strictly dichotomous political choices. Whether or not agenda manipulation is a problem that can be solved, it makes at least theoretical sense to see political choices as ideally reducible to alternatives between two options.

As to the idea that politics is also about identifying, formulating, and occasionally creating the options, rather than choosing among predefined ones, it would be hard not to agree as well. Human beings shape their environment and construct in part the maze in which they evolve, determining the nature of the choices they face at every fork. Human beings carve out in the world options that make sense to them, under the constraints of time, information, and cognitive abilities that characterize them. Again, the metaphor of the maze is meant as a simplifying and clarifying thought experiment. The fact that it cannot account for the complexity of real human choices should not necessarily be held against it.

4. Empirical Segue to the Theoretical Epistemic Claim

The epistemic claim is theoretical, in the sense that I have supported it essentially with abstract arguments (mathematical theorems and models in particular) and only a limited and mostly anecdotal amount of empirical evidence. An interesting next step would be to translate that claim into a testable empirical hypothesis and try to verify or disprove it. This book has tried to provide examples of democratic intelligence, particularly at the micro level (of small-scale democratic assemblies). These examples, however, do not amount to an empirical demonstration of the epistemic strength of democratic decision making.

An interesting and ambitious task at this point would be to support the maximal epistemic argument presented here with empirical claims connecting existing democracies and a number of valued outcomes. Josiah Ober's (2010) case study of ancient Athens provides historical evidence in favor of the epistemic superiority of (direct) democracy over nondemocratic ones. The extent to which this case study can be generalized, and generalized in particular to the case of modern representative democracy, remains an open question. Another interesting question is whether Amartya Sen's hypothesis that democracies do not starve their people, even in countries where crop failures and other natural disasters are common (Sen 1999: chaps. 6 and 7), can be turned into an epistemic claim. While Sen proposes a Schumpeterian account of that empirical correlation between democratic rule and avoidance of famines, defending the idea that the holding of periodic elections, combined with a free press and a certain degree of political transparency ensure that leaders do not take the risk of letting their people go hungry, one could also propose an account of the avoidance of famines in terms of democratic reason. Another example of empirical claims that could be connected with the hypothesis of democratic reason is the attempt to show that democracy actually fosters economic growth. While the evidence for that causal claim is debated (see, e.g., Przeworski 2000),[41] there are comforting correlations between democratic forms of governance and growth indicators. To quote the conclusion of at least one study that is critical

---

[41] According to this impressive contribution to the statistical literature on democracy and growth, while there is no proof that democracy need be sacrificed on the altar of development, "the recently heralded economic virtues of democracy are yet another figment of the ideological imagination" (Przeworski 2000: 271). In brief, democracy might be correlated with economic prosperity, but that has probably more to do with the fact that democracies survive only under favorable economic circumstances than the fact that they actually generate those favorable economic circumstances.

of skeptics and supportive of the optimistic side in a way that exactly squares with my own argument:

> The average democracy has been better for economic growth than the average autocracy, at least in the formative years before World War II. Perhaps Lee Kuan Yew is the exception that proves the bad-autocracy rule he tried to deny. When it comes to wrecking economies, the tyranny of the majority over the voting minority, popularized by our reading De Tocqueville, may have been no match for the tyranny over those with no voice at all. (Lindert 2003: 344)

As to the evidence that would count against the epistemic claim, none of the classic examples of democratic failures—the Sicily expedition, Hitler's election, and so on—or any other local, anecdotal examples are in and of themselves capable of refuting a claim that is admittedly probabilistic. One or even a few black swans—or Lee Kuan Yew—do not falsify the claim that democracy is, on average and in the long run, the smarter rule. What would be needed in order to refute that claim is a demonstration that democracies do systematically worse than competing regimes. This demonstration is lacking. The recent and most promising attempt at showing that democracies "choose bad policies" (Caplan 2007) does manage to show that democracies often choose suboptimal economic policies compared with what an elite of economists would settle for. This is a far cry from the radical debunking promised by the subtitle, but it still suggests that democracies left to their own devices should do worse than, say, countries following the advice of the World Bank (especially in the heyday of the Washington Consensus). I will let the reader judge whether this assertion is, today, more empirically supported than the reverse epistemic claim in favor of democratic reason.[42]

5. THE WISDOM OF THE PAST MANY AND DEMOCRACY AS A LEARNING PROCESS

The stress on collective intelligence is not meant to deny, however, the seriousness of the risks of "democratic unreason." I have mentioned the importance of introducing or maintaining the conditions conducive to the expression of cognitive diversity in order to avoid the dangers induced by deliberation among the like-minded and the tyranny of conformist

---

[42] And for a little help, see "Ending Famines, Simply by Ignoring the Experts," *New York Times*, December 2, 2007, http://www.nytimes.com/2007/12/02/world/africa/02malawi.html, accessed December 10, 2007.

majorities. Being vigilant about cognitive diversity, however, might not be enough. Some choices involve risks that are too serious to be taken in a purely democratic fashion—that is, deliberation followed by simple majority rule. These risks are connected to the abuse of minorities. This is where history and the temporal dimension of democratic reason become a crucial part of the story—one that this book had little opportunity to explore, but that I would like to mention in this conclusion as a potential field of investigation for further research.

A crucial element of intelligence is the ability to learn from one's mistakes. This in turn requires the ability to store and process knowledge's of one's past. What is true of individuals is true of societies and regimes as well. The concept of democratic reason must thus include the ability of a people to remember the past and learn from it, which implies that democracies must have some specific institutions designed to "store" that memory and that knowledge. In other words, democratic reason must not just be distributed across space but across time as well. Institutions that serve as "cognitive artifacts" containing the knowledge of past generations, in the same way as a to-do list contains the knowledge of a past self, might ensure that deliberation and majority rule produce decisions in keeping with the lessons learnt from the past. The most obvious candidate for such a function of collective memory is a constitution, but museums and other cultural institutions can be seen as performing a similar function in an indirectly political way. The lessons contained in those collective artifacts might dictate an occasional limitation of free speech (by banning hate speech, say) or decision making by supermajorities (in effect, the rule of the few).

The idea of collective cognitive artifacts containing the wisdom of the past has conservative roots. Burke famously saw social norms and traditions as expressing such wisdom of the past and forming for that reason a surer guide to action than the instantaneous calculus of reason. But we saw how a radical philosopher like Dewey also suggested that what he called "social intelligence" was in part "embodied intelligence"—the knowledge contributed by many people, in however small amounts, and crystallized and fixed for the future in public knowledge. Further, the use of and respect for the past need not be interpreted in the sense of a dictatorship of the dead. One could imagine a Jeffersonian scheme, in which the content of the Constitution would be renegotiated every twenty-five years or so in order to make it more democratic, but which would still retain the moderating function of constitutional principles as an essential element of the political landscape. The point, in any case, is that where the power of cognitive diversity and democratic norms fail, the wisdom of the "past many" contained in an (ideal) constitution can constrain democratic deliberation or majority rule in ways that can keep us safe from democratic unreason.

6. REASON AND RATIONALITY

The approach in terms of collective intelligence put forward by this book finally allows me to raise a question regarding the relevance of a lot of research on "public opinion." What my approach suggests is that the research on the uninformed voter is deeply misguided. Requiring of the average citizen that he be able to answer questions about the proportion of Congress it takes to overturn a presidential veto is, to put it metaphorically, like asking someone to compute a complex multiplication problem without a calculator or the required cognitive artifact. The fact that people cannot answer the question does not show that they are stupid or incompetent or incapable of ruling. It just goes to show that the cognitive unit that is of interest is bigger than the individual and probably involves other people and a number of cognitive artifacts. Deliberation and the technique of voting are such cognitive artifacts, allowing us to distribute the cognitive burden and reason collectively. As suggested earlier, a constitution or the institution of judicial review might be another way to distribute the task of finding a good answer to a public problem, this time by distributing the social calculus not just through space, but through time, across past generations. Collective distributed intelligence in politics may thus span a vast array of institutions, all of which might be needed to constrain, guide, and even correct citizens' individual mental operations.

The notion of distributed intelligence thus raises interesting conceptual problems when considered in relation to that of "rationality." Rationality is perhaps a more tractable notion than intelligence, but its poverty is also probably why political sciences keep running into "paradoxes" at both the individual and collective levels. This book thus suggests a more systemic approach to political decision making, one that does not focus so much on the average citizen's rationality and information level as the only explanatory variable for the quality of democratic outcomes, but "distributes" the explanation onto the whole of the democratic system.

On the normative level, this also has implications. If collective intelligence is a real phenomenon, then we should worry less about the failure of, for example, educational programs to improve individuals' cognitive abilities, and worry more about their ability to enhance the intelligence of the democratic system as a whole.

Bibliography

Aboulafia, Mitchell, Myra Bookman, and Catherine Kemp. 2002. *Habermas and Pragmatism*. New York: Routledge.

Abramowicz, Michael. 2007. *Predictocracy: Market Mechanisms for Public and Private Decision Making*. New Haven, CT: Yale University Press.

Abramson, Jeffrey. (1994) 2001. *We the Jury*. Cambridge, MA: Harvard University Press.

Ackerman, Bruce, and James S. Fishkin. 2004. *Deliberation Day*. New Haven, CT: Yale University Press.

Adams, John. (1851) 1856. "Thoughts on Government." In *The Works of John Adams, Second President of the United States: With a Life of the Author, Notes and Illustrations, by His Grandson Charles Francis Adams*, vol. 4. Boston: Little, Brown. E-book available at http://oll.libertyfund.org/title/2102.

Althaus, Scott. 2003. *Collective Preferences in Democratic Politics*. Cambridge: Cambridge University Press.

Anderson, Elizabeth. 2006. "The Epistemology of Democracy." *Episteme: A Journal of Social Epistemology*, 3, no. 1–2: 8–22.

Arendt, Hannah. 1993. *Between Past and Future: Eight Exercises in Political Thought*. New York: Penguin Books, 227–64.

———. 2005. *The Promise of Politics*. New York: Shocken Books.

Aristotle. 1991. *On Rhetoric: A Theory of Civic Discourse*. Translated by G. Kennedy. Oxford: Oxford University Press.

———. 1998. *Politics*. Translated by C.D.C. Reeve. Indianapolis: Hackett Publishing.

Arrow, Kenneth J. 1953. *Social Choice and Individual Values*. 2d ed. New York: Wiley, 1963.

Arrow, Kenneth J., Robert Forsythe, Michael Gorham, Robert Hahn, Robin Hanson, John O. Ledyard, Saul Levmore, Robert Litan, Paul Milgrom, Forrest D. Nelson, George R. Neumann, Marco Ottaviani, Thomas C. Schelling, Robert J. Shiller, Vernon L. Smith, Erik Snowberg, Cass R. Sunstein, Paul C. Tetlock, Philip E. Tetlock, Hal R. Varian, Justin Wolfers, and Eric Zitzewitz. 2008. "The Promise of Prediction Markets." *Science* 320, no. 5878: 877–78, DOI:10.1126/science.1157679.

Asad, Muhammad. 1980. *The Principles of State and Government in Islam*. Gibraltar: Dar al-Andalus; Ann Arbor, MI: New Era.

Austen-Smith, David, and Jeffrey S. Banks. 1996. "Information Aggregation, Rationality and the Condorcet Jury Theorem." *American Political Science Review* 90, no. 1: 34–45.

Austen-Smith, David, and Timothy J. Feddersen. 2006. "Deliberation, Preference Uncertainty, and Voting Rules." *American Political Science Review* 100: 209–18.

Bächtiger, Andre, Simon Niemeyer, Michael Neblo, Marco R. Steenbergen, and Jürg Steiner. 2010. "Disentangling Diversity in Deliberative Democracy Competing Theories, Their Blind Spots and Complementarities." *Journal of Political Philosophy* 18, no. 1: 32–63.

Baker, Keith M. 1976. *Condorcet: Selected Writings*. Indianapolis: Bobbs-Merrill.

———. 1982. *Condorcet: From Natural Philosophy to Social Mathematics*. Chicago: University of Chicago Press.

Barabas, Jason. 2000. "Uncertainty and Ambivalence in Deliberative Opinion Models: Citizens in the Americans Discuss Social Security Forum." Prepared for Annual Meeting of the Midwest Political Science Association, Chicago.

Barber, Benjamin. 1985. *Strong Democracy: Participatory Politics for a New Age*. Berkeley: University of California Press.

Barkow, Jerome H., Leda Cosmides, and John Tooby. 1992. *The Adapted Mind*. Oxford: Oxford University Press.

Barry, Brian. (1965) 1990. *Political Argument: A Reissue with a New Introduction*. Berkeley: University of California Press.

Bartels, Larry. 1996. "Uninformed Votes: Information Effects in Presidential Elections." *American Journal of Political Science* 40: 194–230.

Beitz, Charles R. 1990. *Political Equality*. Princeton, NJ: Princeton University Press.

Benhabib, Seyla. 1996. "Toward a Deliberative Model of Democratic Legitimacy." In *Democracy and Difference: Contesting the Boundaries of the Political*, edited by S. Benhabib, 67–94. Princeton, NJ: Princeton University Press.

Berelson, Bernard R., Paul F. Lazarsfeld, and William N. McPhee. 1954. *Voting: A Study of Opinion Formation in a Presidential Campaign*. Chicago: University of Chicago Press.

Bernstein, Richard J. 1992. *The New Constellation: The Ethical-Political Horizons of Modernity/Postmodernity*. Cambridge, MA: MIT Press.

Bessette, Joseph. 1980. "Deliberative Democracy: The Majority Principle in Republican Government." In *How Democratic Is the Constitution?* edited by Robert Goldwin and William Shambra, 102–16. Washington, DC: American Enterprise Institute.

Besson, Samantha, and José Luis. Martí. 2006. *Deliberative Democracy and Its Discontents*. Burlington, VT: Ashgate.

Besson, Samantha, and José Luis Martí, eds. 2007. *Democracy and Its Discontents: National and Post-national Challenges*. Burlington, VT: Ashgate.

Billig, Michael. 1996. *Arguing and Thinking: A Rhetorical Strategy to Social Psychology*. Cambridge: Cambridge University Press.

Black, Duncan. 1958. *Theory of Elections and Committees*. Cambridge: Cambridge University Press.

Bohman, James. 1998. "Survey Article: The Coming of Age of Deliberative Democracy." *Journal of Political Philosophy* 6, no. 4: 400–425.

———. 2006. "Deliberative Democracy and the Epistemic Benefits of Diversity." *Episteme* 3: 175–91.

———. 2007. "Political Communication and the Epistemic Value of Diversity: Deliberation and Legitimation in Media Societies." *Communication Theory* 17, no. 4: 348–55.

Bohman, James, and William Regh, eds. 1997. *Deliberative Democracy: Essays on Reason and Politics*. Cambridge, MA: MIT Press.

Boring, Edwin G. 1923. "Intelligence as the Tests Test It." *New Republic* 36 (1923): 35–37.

Bovens, Luc, and Wlodek Rabinowicz. 2006. "Democratic Answers to Complex Questions—An Epistemic Perspective." *Synthese* 150, no. 1: 131–53.

Bowles, Samuel, and Herbert Gintis. 2006. "Homo Economicus and Zoon Politikon: Behavioral Game Theory and Political Behavior." *Oxford Handbook of Contextual Political Analysis*, edited by R. Goodin and C. Tilly, 176–82. Oxford: Oxford University Press.

Brennan, Geoffrey. 2001. "Collective Coherence?" *International Review of Law and Economics* 21: 197–211.

Brennan, Geoffrey, and Alan Hamlin. 1998. "Expressive Voting and Electoral Equilibrium." *Public Choice* 95: 149–75.

Brennan, Geoffrey, and Loren Lomasky. 1993. *Democracy and Decision: The Pure Theory of Electoral Preference*. Cambridge: Cambridge University Press.

Brennan, Jason. 2011. *The Ethics of Voting*. Princeton, NJ: Princeton University Press.

Breton, Philippe. 2006. *L'incompétence démocratique: La crise de la parole aux sources du malaise (dans la) politique*. Paris: La Découverte.

Bull, Malcolm. 2005. "The Limits of Multitude." *New Left Review* 35 (September/October): 19–39.

Burnett, Craig, and Matthew McCubbins. 2010. "What Do You Know? Comparing Political and Consumer Knowledge." Unpublished paper. Available at http://ssrn.com/abstract=1493533.

Burns, James Henderson. 1957. "J. S. Mill and Democracy, 1829–61: II." *Political Studies* 5, no. 3: 281–94.

Callenbach, Earnest, and Michael Phillips. 1985. *A Citizen Legislature*. Berkeley, CA: Banyan Tree Books/Clear Glass.

Callon, Michel, Pierre Lascoumes, and Yannick Barthe. 2001. *Agir dans un monde incertain: Essai sur la démocratie technique*. Paris: Seuil.

Caplan, Bryan. 2007. *The Myth of the Rational Voter: Why Democracies Choose Bad Policies*. Princeton, NJ: Princeton University Press.

Carson, Lyn, and Brian Martin. 1999. *Random Selection in Politics. (Luck of the Draw: Sortition and Public Policy)*. Westport, CT: Praeger Publishers.

Chapman, Bruce. 1998. "More Easily Done than Said: Rules, Reason and Rational Social Choice." *Oxford Journal of Legal Studies* 18: 293–329.

Christiano, Thomas. 1997. "The Significance of Public Deliberation." In *Deliberative Democracy: Essays on Reason and Politics*, edited by J. Bohman and W. Regh, 243–78. Cambridge, MA: MIT Press.

———. 1996. *The Rule of the Many*. Boulder, CO: Westview Press.

———. 2000. "A Democratic Paradox?" *Political Science Quarterly* 115, no. 1: 35–40.

Clark, Andy. 1998. *Being There: Putting Brain, Body, and World Together Again*. Cambridge, MA: MIT Press.

Cohen, Gerald A. 2003. "Facts and Principles." *Philosophy and Public Affairs* 31, no. 3: 211–45.

Cohen, Joshua. 1986. "An Epistemic Conception of Democracy." *Ethics* 97, no. 1: 26–38.

———. 1989. "Deliberation and Democratic Legitimacy." In *The Good Polity*, edited by A. Hamlin and P. Pettit, 17–34. New York: Basil Blackwell.

———. 1996. "Procedure and Substance in Deliberative Democracy." In *Democracy and Difference: Contesting the Boundaries of the Political*, edited by S. Benhabib, 95–119. Princeton, NJ: Princeton University Press.

———. 2004. "Social Choice." Unpublished paper. Available at http://ocw. mit.edu/NR/rdonlyres/Political-Science/17-960Fall-2004/9B1453C3-37F0 -47A1-A296-B496AF244A09/0/social_lecnote.pdf (accessed June 14, 2006).

———. 2009. "Truth and Public Reason." *Philosophy and Public Affairs* 37, no. 1: 2–42.

Coleman, Jules, and John Ferejohn. 1986. "Democracy and Social Choice." *Ethics* 97, no. 1: 6–25.

Coleman, Stephen, and Jay G. Blumler. 2009. *The Internet and Democratic Citizenship: Theory, Practice, and Policy*. Cambridge: Cambridge University Press.

Condorcet, Marie Jean Antoine Nicolas de. 1785. *Essai sur l'application de l'analyse à la probabilité des décisions rendues à la pluralité des voix*. Paris: De l'Imprimerie Royale.

———. 1968. *Oeuvres: Nouvelle impression en fac-similé de l'édition (Paris 1847–1849)*. Edited by M. F. Arago and A. Condorcet-O'Connor. 12 vols. Stuttgard-Bad Cannstatt: Friedrich Frommann.

———. 1976. *Condorcet: Selected Writings*. Edited by Michael K. Baker. Indianapolis: Bobbs-Merrill.

Conradt, Larissa, and Timothy J. Roper. 2007. "Democracy in Animals: The Evolution of Shared Group Decisions." *Proceedings of the Royal Society: Biological Sciences* 274, no. 1623: 2317.

Converse, Philip E. 1964. "The Nature of Belief Systems in Mass Publics." In *Ideology and Discontent*, edited by D. E. Apter, 206–61. New York: Free Press.

———. 1990. "Popular Representation and the Distribution of Information." In *Information and Democratic Processes*, edited by J. A. Ferejohn and J. H. Kuklinski, 369–89. Chicago: University of Illinois Press.

———. 2000. "Assessing the Capacity of Mass Electorates." *Annual Review of Political Science* 3: 331–53.

Cook, Faye Lomax, and Lawrence R. Jacobs. 1998. *Deliberative Democracy in Action: Evaluation of Americans Discuss Social Security*. Washington, DC: Report to the Pew Charitable Trusts.

Cooke, Maeve. 2000. "Five Arguments for Deliberative Democracy." *Political Studies* 48: 947–69.

Copp, David, Jean Hampton, and John E. Roemer, eds. 1993. *The Idea of Democracy*. Cambridge: Cambridge University Press.

Cosmides, L., and J. Tooby. 1992. "Cognitive Adaptations for Social Exchange." In *The Adapted Mind: Evolutionary Psychology and the Generation of Culture*, edited by J. H. Barkow, L. Cosmides, and J. Tooby, 163–228. New York: Oxford University Press.

Crépel, Pierre, and Christian Gilain (sous la direction de). *Colloque international. Condorcet: mathématicien, économiste, philosophe, homme politique.* Paris: Minerve.

Dahl, Robert. 1989. *Democracy and Its Critics.* New Haven, CT: Yale University Press.

Dekel, Eddie, and Michele Piccione. 2000. "Sequential Voting Procedures in Symmetric Binary Elections." *Journal of Political Economy* 108, no. 1: 34–55.

Delli Carpini, Michael X., Faye Lomax Cook, and Lawrence R. Jacobs. 2004. "Public Deliberation, Discursive Participation, and Citizen Engagement: A Review of the Empirical Literature." *Annual Review in Political Science* 7: 315–44.

Delli Carpini, Michael X., and Scott Keeter. 1996. *What Americans Know about Politics and Why It Matters.* New Haven, CT: Yale University Press.

Deneen, Patrick. 2005. *Democratic Faith.* Princeton, NJ: Princeton University Press.

Denes-Raj, Veronika, and Seymour Epstein. 1994. "Conflict between Intuitive and Rational Processing: When People Behave against Their Better Judgment." *Journal of Personality and Social Psychology* 66, no. 5: 819–29.

Dennett, Daniel C. 1987. *The Intentional Stance.* Cambridge, MA: MIT Press.

Dewey, John. 1916. *Democracy and Education.* E-book available at http://www.netlibrary.com.ezp1.harvard.edu/Reader/.

———. (1927) 1954. *The Public and Its Problems.* Chicago: Swallow Press.

Dietrich, Franz, and Christian List. 2004. "A Model of Jury Decisions where All Jurors Have the Same Evidence." *Synthese* 142, no. 2: 175–202.

———. 2007. "Arrow's Theorem in Judgment Aggregation." *Social Choice and Welfare* 29, no. 1: 19–33.

Dijksterhuis, Ap. 2004. "Think Different: The Merits of Unconscious Thought in Preference Development and Decision Making." *Journal of Personality and Social Psychology* 87: 586–98.

Downs, Anthony. 1957. *An Economic Theory of Democracy.* New York: Harper and Row.

Dryzek, John S. 2000. *Deliberative Democracy and Beyond: Liberals, Critics, Contestations.* New York: Oxford University Press.

Dryzek, John S., and Simon Niemeyer. 2006. "Reconciling Pluralism and Consensus as Political Ideals." *American Journal of Political Science* 50, no. 3: 634–49.

Dunn, John. 2005. *Setting the People Free: The Story of Democracy.* London: Atlantic Books.

Duxbury, Neil. 1999. *Random Justice: On Lotteries and Legal Decision-Making.* Oxford: Oxford University Press.

Dworkin, Ronald. 1996. "Objectivity and Truth: You'd Better Believe It." *Philosophy and Public Affairs* 25, no. 2: 87–139.

Elkin, Stephen L., and Karol Edward Soltan. 1999. *Citizen Competence and Democratic Institutions.* University Park, PA: Pennsylvania State University Press.

Elster, Jon. 1986. *Foundations of Social Choice Theory.* Cambridge: Cambridge University Press.

———. 1989a. "The Market and the Forum: Three Varieties of Political Theory." In *Foundations of Social Choice Theory*, edited by J. Elster and A. Hylland, 104–32. Cambridge: Cambridge University Press.

———. 1989b. *Solomonic Judgments: Studies in the Limits of Rationality.* Cambridge: Cambridge University Press; Paris: Editions de la Maison des sciences de l'homme.

———. 1995. "Strategic Uses of Argument." In *Barriers to Conflict Resolution*, edited by Kenneth Arrow, Robert Mnookin, Lee Ross, Amos Tversky, and Robert Wilson, 237-257. New York: Norton.

———, ed. 1998. *Deliberative Democracy.* Cambridge: Cambridge University Press.

———. 2000. *Ulysses Unbound: Studies in Rationality, Precommitment, and Constraints.* Cambridge: Cambridge University Press.

———. 2007. *Explaining Social Behavior: More Nuts and Bolts for the Social Sciences.* Cambridge: Cambridge University Press.

———. 2009. *Le désintéressement: Critique de l'homme économique.* Paris: Le Seuil.

Elster, Jon, and Hélène Landemore. 2008. "Ideology and Dystopia." *Critical Review* 20, no. 3: 273–89.

Engelstad, Fredrik. 2011. "Lot as a Democratic Device of Selection." In *Lotteries in Public Life. A Reader*, edited by P. Stone, 177–200. Exeter, UK: Imprint Academic.

Ermakoff, Ivan. 2008. *Ruling Oneself Out: A Theory of Collective Abdications.* Durham, NC: Duke University Press.

Estlund, David. 1989. "Democracy and the Common Interest: Condorcet and Rousseau Revisited." *American Political Science Review 83*: 1317–22.

———. 1993a. "Making Truth Safe for Democracy." In *The Idea of Democracy*, edited by D. Copp, J. Hampton, and J. Roemer, 71–100. Cambridge: Cambridge University Press.

———. 1993b. "Who's Afraid of Deliberative Democracy? On the Strategic/Deliberative Dichotomy in Recent Constitutional Jurisprudence." *Texas Law Review* 71: 1437–77.

———. 1994. "Opinion Leaders, Independence, and Condorcet's Jury Theorem." *Theory and Decision* 36, no. 2: 131–62.

———. 1997. "Beyond Fairness and Deliberation: The Epistemic Dimension of Democratic Authority." In *Deliberative Democracy: Essays on Reason and Politics*, edited by J. Bohman and W. Rehg, 173–204. Cambridge, MA: MIT Press.

———. 1998. "The Insularity of the Reasonable: Why Political Liberalism Must Admit the Truth." *Ethics* 108, no. 3: 252–75.

———, ed. 2002. *Democracy.* Malden, MA: Blackwell.

———. 2008. *Democratic Authority: A Philosophical Framework.* Princeton, NJ: Princeton University Press.

Estlund, David, and Robert E. Goodin. 2004. "The Persuasiveness of Democratic Majorities." *Politics, Philosophy and Economics* 3, no. 2: 131–42.

Evans, Jonathan S. "Logic and Human Reasoning: An Assessment of the Deduction Paradigm." *Psychological Bulletin* 128, no. 6: 978–96.

Evans, Jonathan S., and David Over. 1996. *Rationality and Reasoning*. Hove, UK: Psychology Press.

Farrar, Cynthia. 1988. *The Origins of Democratic Thinking: The Invention of Politics in Classical Athens*. Cambridge: Cambridge University Press.

Farrar, Cynthia, James Fishkin, Donald Green, Christian List, Robert Luskin, Elizabeth Paluck, 2010. "Disaggregating Deliberation's Effects: An Experiment within a Deliberative Poll." *British Journal of Political Science*, 40, no. 2: 333–47.

Fauré, Christine. 1989. "La pensée probabiliste de Condorcet et le suffrage féminin." In *Colloque international: Condorcet—mathématicien, économiste, philosophe, homme politique*, under the direction of Pierre Crépel and Christian Gilain, 349–54. Paris: Minerve.

Feddersen, Timothy, and Wolfgang Pesendorfer. 1997. "Voting Behavior and Information Aggregation in Elections with Private Information." *Econometrica* 65, no. 5: 1029–58.

———. 1998. "Convicting the Innocent: The Inferiority of Unanimous Jury Verdicts under Strategic Voting." *American Political Science Review* 92, no. 1: 23–35.

———. 1999. "Elections, Information Aggregation, and Strategic Voting." *Proceedings of the National Academy of Sciences of the United States of America* 96, no. 19: 10572.

Ferejohn, John. 2005. "The Citizens' Assembly Model." Unpublished manuscript. Available at http://www.hss.caltech.edu/media/seminar-papers/ferejohn.pdf.

Ferejohn, John A., and James H. Kuklinski. 1990. *Information and Democratic Processes*. Urbana: University of Illinois Press.

Finley, Moses I. (1973) 1985. *Democracy Ancient and Modern*. New Brunswick, NJ: Rutgers University Press.

Fishkin, James. 1991. *Democracy and Deliberation*. New Haven, CT: Yale University Press.

———. 1995. *The Voice of the People: Public Opinion and Democracy*. New Haven, CT: Yale University Press.

———. 2009. *When the People Speak: Deliberative Democracy and Public Consultation*. Oxford: Oxford University Press.

Fishkin, James, and Bruce Ackerman. 2004. *Deliberation Day*. New Haven, CT: Yale University Press.

Fishkin, James S., and R. C. Luskin. 2005. "Experimenting with a Democratic Ideal: Deliberative Polling and Public Opinion." *Acta Politica* 40, no. 3: 284–98.

Frederick, Shane. 2005. "Cognitive Reflection and Decision Making." *Journal of Economic Perspectives* 19, no. 4: 25–42.

Fuller, Steve. 1987. "On Regulating What Is Known: A Way to Social Epistemology." *Synthese* 73, no. 1: 145–83.

———. 1988. *Social Epistemology*. Bloomington: Indiana University Press.

———. 1993. *Philosophy, Rhetoric, and the End of Knowledge*. Madison: University of Wisconsin Press.

———. 1999. *The Governance of Science: Ideology and the Future of the Open Society*. London: Open University Press.

Fung, Archon, and Erik Olin Wright, eds. 2003. *Deepening Democracy: Institutional Innovations in Empowered Participatory Governance*. London: Verso.

Galston, William A. 2001. "Political Knowledge, Political Engagement, and Civic Education." *Annual Review of Political Science* 4: 217–34.

Galton, Francis. 1907a. "The Ballot-Box." *Nature* 75 (March 28): 509–10.

———. 1907b. "Vox Populi." *Nature* 75 (March 7): 450–51.

———. 1908. *Memories of My Life*. London: Methuen.

Gardner, Howard. 1983. *Frames of Mind: The Theory of Multiple Intelligences*. New York: Basic.

Gastil, John, and James P. Dillard. 1999. "Increasing Political Sophistication through Public Deliberation." *Political Communication* 16, no. 1: 3–23.

Gaus, Gerald. 1996. *Justificatory Liberalism: An Essay on Epistemology and Political Theory*. Oxford: Oxford University Press.

———. 1997a. "Looking for the Best and Finding None Better: The Epistemic Case for Deliberative Democracy." *Modern Schoolman* 74: 277–84.

———. 1997b. "Reason, Justification, and Consensus: Why Democracy Can't Have It All." In *Deliberative Democracy: Essays on Reason and Politics*, edited by J. Bohman and W. Regh, 205–42. Cambridge, MA: MIT Press.

Gehrlein, William V. 2002. "Condorcet's Paradox and the Likelihood of Its Occurrence: Different Perspectives on Balanced Preferences." *Theory and Decision* 52: 171–99.

Gibbard, Allan. 1990. *Wise Choices, Apt Feelings: A Theory of Normative Judgment*. Cambridge, MA: Harvard University Press.

Gigerenzer, Gerd, Peter M. Todd, and the ABC Research Group. 2000. *Simple Heuristics That Make Us Smart*. Oxford: Oxford University Press.

Gigerenzer, Gerd, and Rheinard Selten. 2001. *Bounded Rationality: The Adaptive Toolbox*. Cambridge, MA: MIT Press.

Goldman, Alvin. 1986. *Epistemology and Cognition*. Cambridge, MA: Harvard University Press.

———. 1987. "Foundations of Social Epistemics." *Synthese* 73: 109–44.

———. 1999. *Knowledge in a Social World*. Oxford: Oxford University Press.

———. 2004. "Group Knowledge versus Group Rationality: Two Approaches to Social Epistemology." *Episteme* 1, no. 1: 11–22.

Goodin, Robert E. 1986. "Laundering Preferences." In *Foundations of Social Choice Theory*, edited by J. Elster and A. Hylland, 75–102. Cambridge: Cambridge University Press.

———. 2000. "Democratic Deliberation Within." *Philosophy and Public Affairs* 29, no. 1: 81–109.

———. 2003. *Reflective Democracy*. Oxford: Oxford University Press.

———. 2008. *Innovating Democracy: Democratic Theory and Practice after the Deliberative Turn*. Oxford: Oxford University Press.

Goodin, Robert E., and Christian List. 2001. "Epistemic Democracy: Generalizing the Condorcet Jury Theorem." *Journal of Political Philosophy* 9, no. 3: 277–306.

———. 2006. "A Conditional Defense of Plurality Rule: Generalizing May's Theorem in a Restricted Informational Environment." *American Journal of Political Science* 50, no. 4: 940–49.

Goodin, Robert E., and Simon J. Niemeyer. 2003. "When Does Deliberation Begin? Internal Reflection versus Public Discussion in Deliberative Democracy." *Political Studies* 51, no. 4: 627–49.

Goodin, Robert E., and Kevin W. S. Roberts. 1975. "The Ethical Voter." *American Political Science Review* 69: 926–28.

Goodwin, Barbara. 1992. *Justice by Lottery*. Chicago: University of Chicago Press.

Gottfredson, Linda S., ed. 1997. "Foreword to 'Intelligence and Social Policy.' " *Intelligence* 24, no. 1 (special issue): 1–12.

Gould, Carol C. 1988. *Rethinking Democracy: Freedom and Social Cooperation in Politics, Economy, and Society*. New York: Cambridge University Press.

Greenberg, Edward S. 1986. *Workplace Democracy: The Political Effects of Participation*. Ithaca, NY: Cornell University Press.

Grofman, Bernard, and Scott L. Feld. 1988. "Rousseau's General Will: A Condorcetian Perspective." *American Political Science Review* 82: 567–76.

Grofman, Bernard, and Guillermo Owen, eds. 1986. *Information Pooling and Group Decision Making*. Greenwich, CT: JAI.

Grofman, Bernard, Guillermo Owen, and Scott L. Feld. 1983. "Thirteen Theorems in Search of Truth." *Theory and Decision* 15: 261–78.

Grunberg, Gérard. 2002. "Le soutien à la démocratie représentative." In *La Démocratie à l'épreuve: Une nouvelle approche de l'opinion des Français*, under the direction of G. Grunberg, N. Mayer, and P. M. Sniderman. Paris: Presses de Sciences Po.

Guenther, Corey L., and Mark D. Alicke. 2008. "Self-Enhancement and Belief Perseverance." *Journal of Experimental Social Psychology* 44, no. 3: 706–12.

Gutmann, Amy, and Dennis Thompson. 1996. *Democracy and Disagreement*. Cambridge, MA: Harvard University Press.

———. 2002. "Deliberative Democracy beyond Process." *Journal of Political Philosophy* 10, no. 2: 153–74.

———. 2004. *Why Deliberative Democracy?* Princeton, NJ: Princeton University Press.

Habermas, Jürgen. 1981. *Theorie des Kommunikativen Handelns*, vol. 1. Frankfurt: Suhrkamp.

———. (1977) 1984. *The Theory of Communicative Action*. Vol. 1, *Reason and the Rationalization of Society*, translated by Thomas McCarthy. Boston: Beacon Press.

———. 1990a. "The Division of Cognitive Labor." *Journal of Philosophy* 87: 5–22.

———. 1990b. *Moral Consciousness and Communicative Action*. Cambridge, MA: MIT Press.

———. 1993. *The Advancement of Science*. Oxford: Oxford University Press.

———. 1995. "Reconciliation through the Public Use of Reason: Remarks on John Rawls's Political Liberalism." *Journal of Moral Philosophy* 92, no. 3: 109–31.

———. 1996. *Between Facts and Norms: Contributions to a Discourse Theory of Law and Democracy*. Translated by W. Rehg. Cambridge: Polity Press.

——. 1997. "Popular Sovereignty as Procedure." In *Deliberative Democracy: Essays on Reason and Politics*, edited by J. Bohman and W. Rehg, 35–65. Cambridge, MA: MIT Press.

——. 2003. "Rightness versus Truth: On the Sense of Normative Validity in Moral Judgments and Norms." In J. Habermas, *Truth and Justification*, translated by B. Fultner, 237–75. Cambridge, MA: MIT Press.

——. 2004. *Wahrheit und Rechfertigung*. Suhrkamp: Frankfurt.

——. 2006. "Political Communication in Media Society: Does Democracy Still Enjoy an Epistemic Dimension? The Impact of Normative Theory on Empirical Research." *Communication Theory* 16, no. 4: 411–26.

——. 2008. *Between Naturalism and Religion: Philosophical Essays*. Translated by Ciaran Cronin. Cambridge: Polity Press.

Hacking, Ian. 1984. *The Emergence of Probability*. Cambridge, MA: Cambridge University Press.

——. 1990. *The Taming of Chance*. Cambridge: Cambridge University Press.

Hanson, Robin. 2007. "Shall We Vote on Values, but Bet on Beliefs?" available at http://hanson.gmu.edu/futarchy.pdf.

Hardin, Russell. 1980. "Rationality, Irrationality and Functionalist Explanation." *Social Science Information* 19: 755–72.

——. 1999. *Liberalism, Constitutionalism, and Democracy*. Oxford: Oxford University Press.

Hayek, Friedrich. 1945. "The Use of Knowledge in Society." *American Economic Review* 35: 519–30.

——. 1982. *Law, Legislation, and Liberty: A New Statement of the Liberal Principles of Justice and Political Economy*. London: Routledge.

Heath, Joseph. 2001. *Communicative Action and Rational Choice*. Cambridge, MA: MIT Press.

Hibbing, John R., and Elizabeth Theiss-Morse. 1995. *Congress as Public Enemy: Public Attitudes toward American Political Institutions*. New York: Cambridge University Press.

Hong, Lu, and Scott Page. 2001. "Problem Solving by Heterogeneous Agents." *Journal of Economic Theory* 97, no. 1: 123–63.

——. 2004. "Groups of Diverse Problem Solvers Can Outperform Groups of High-Ability Problem Solvers." *Proceedings of the National Academy of Sciences of the United States* 101, no. 46: 16385–89.

——. 2009. "Interpreted and Generated Signals." *Journal of Economic Theory* 144, no. 5: 2174–96.

——. 2012. "Micro-foundations of Collective Wisdom." In *Collective Wisdom: Principles and Mechanisms*, edited by H. Landemore and J. Elster. Cambridge: Cambridge University Press.

Hutchins, Edwin. 1995. *Cognition in the Wild*. Cambridge, MA: MIT Press.

Jones, W. H. Morris. 1954. "In Defence of Apathy: Some Doubts on the Duty to Vote." *Political Studies* 2, no. 1: 25–37.

Kahneman, Daniel. 2003. "A Perspective on Judgment and Choice: Mapping Bounded Rationality." *American Psychologist* 58, no. 9: 697–720.

Kahneman, Daniel, Paul Slovic, and Amos Tversky. 1982. *Judgment under Uncertainty: Heuristics and Biases*. New York: Cambridge University Press.

Kahneman, Daniel, and Amos Tversky. 1973. "On the psychology of prediction." *Psychological Review* 80: 237–51.

Karotkin, Drora, and Jacob Paroush. 2003. "Optimum Committee Size: Quality-versus-Quantity Dilemma." *Social Choice and Welfare* 20, no. 3: 429–41.

Kerr, Norbert L., Robert J. MacCoun, and Geoffrey P. Kramer. 1996. "Bias in Judgment: Comparing Individuals and Groups." *Psychological Review* 103: 687–719.

Kitcher, Philip. 1990. "The Division of Cognitive Labor." *Journal of Philosophy* 87: 5–22.

———. 1993. *The Advancement of Science*. New York: Oxford University Press.

———. 2001. *Science, Truth, and Democracy*. Oxford: Oxford University Press.

Klarreich, Erica. 2003. "Best Guess: Economists Explore Betting Markets as Prediction Tools." *Science News*, October 18, http://findarticles.com/p/articles/mi_m1200/is_16_164/ai_110459327/?tag=content;col1.

Klein, Stanley, Leda Cosmides, John Tooby, and Sarah Chance. 2002. "Decisions and the Evolution of Memory: Multiple Systems, Multiple Functions." *Psychological Review* 109: 306–32.

Knag, Simon. 2011. "Let's Toss for It: A Surprising Curb on Political Greed." In *Lotteries in Public Life: A Reader*, edited by P. Stone, 251–62. Exeter: Imprint Academic.

Knight, Jack, and James Johnson. 1994. "Aggregation and Deliberation: On the Possibility of Democratic Legitimacy." *Political Theory* 22: 277–96.

Koriat, Asher, Sarah Lichtenstein, and Baruch Fischhoff. 1980. "Reasons for Confidence." *Journal of Experimental Psychology: Human Learning and Memory* 6, no. 2: 107–18.

Kornhauser, Lewis A. 1992. "Modelling Collegial Courts, pt. 2: Legal Doctrine." *Journal of Law, Economics and Organization* 8: 441–70.

Kornhauser, Lewis A., and Larry G. Sager. 2004. "Group Choice in Paradoxical Cases." *Philosophy and Public Affairs* 32: 249–76.

Kuklinski, James H., and Paul J. Quirk. 2000. "Reconsidering the Rational Public: Cognition, Heuristics, and Mass Opinion." In *Elements of Reason*, edited by A. Lupia, M. McCubbins, and S. Popkin, 153–83. Cambridge, UK: Cambridge University Press.

Ladha, Krishna. 1992. "The Condorcet Jury Theorem, Free Speech, and Correlated Votes." *American Journal of Political Science* 36, no. 3: 617–34.

Lamberson, P. J., and Scott Page. Forthcoming. "Optimal Forecasting Groups." *Management Science*.

Landemore, Hélène. 2004. "Politics and the Economist-King: Is Rational Choice Theory the Science of Choice?" *Journal of Moral Philosophy* 1, no. 2: 177–96.

———. 2008. "Is Representative Democracy Really Democratic?" Interview of Bernard Manin and Nadia Urbinati. *La Vie des Idées*, March 7, http://www.laviedesidees.fr/La-democratie-representative-est.html.

———. 2010. "La raison démocratique: Les mécanismes de l'intelligence collective en politique." *Raison Publique* no. 12: 9–55.

———. 2012. "Democratic Reason: The Mechanisms of Collective Intelligence in Politics." In *Collective Wisdom: Principles and Mechanisms*, edited by H. Landemore and J. Elster 251–89. Cambridge: Cambridge University Press.

Landemore, Hélène, and Jon Elster, eds. 2012. *Collective Wisdom: Principles and Mechanisms*. Cambridge: Cambridge University Press.

Landemore, Hélène, and Hugo Mercier. 2010. "Talking It Out: Deliberation with Others vs. Deliberation Within." Paper presented at the American Political Science Annual Convention. Available at http://papers.ssrn.com/sol3/papers.cfm?abstract_id=1660695. Forthcoming with *Análise Social*.

Laughlin, Patrick R., Bryan L. Bonner, and Andrew G. Miner. 2002. "Groups Perform Better than the Best Individuals on Letters-to-Numbers Problems." *Organizational Behavior and Human Decision Processes* 88, no. 2: 605–20.

Laughlin, Patrick, and Alan Ellis. 1986. "Demonstrability and Social Combination Processes on Mathematical Intellective Tasks." *Journal of Experimental Social Psychology* 22: 177–89.

Lave, Jean. 1988. *Cognition in Practice: Mind, Mathematics, and Culture in Everyday Life*. New York: Cambridge University Press.

Lebon, Gustave. (1895) 2006. *La Psychologies des Foules*. Paris: PUF.

Lerner, Jennifer S., and Philip E. Tetlock. 1999. "Accounting for the Effects of Accountability." *Psychological Bulletin* 125: 255–75.

Lindblom, Charles. 1965. *The Intelligence of Democracy: Decision-making through Mutual Adjustment*. New York: Free Press.

Lindert, Peter H. 2003. "Voice and Growth: Was Churchill Right?" *Journal of Economic History* 63, no. 2: 315–50.

Lippincott, Benjamin. (1938) 1964. *Victorian Critics of Democracy: Carlyle, Ruskin, Arnold, Stephen, Maine, Lecky*. New York: Octagon Books.

Lippmann, Walter. (1925) 1993. *The Phantom Public* . New Brunswick, NJ: Transaction Publishers.

Lipset, Seymour M. 1960. *Political Man: The Social Bases of Politics*. New York: Doubleday.

List, Christian. 1998. Mission Impossible? The Problem of Democratic Aggregation in the Face of Arrow's Theorem." DPhil thesis in Politics, University of Oxford.

———. 2001. "Some Remarks on the Probability of Cycles." Published as "Appendix 3: An Implication of the k-Option Condorcet Jury Mechanism for the Probability of Cycles." In List and Goodin 2001.

———. 2002. "Aggregating Sets of Judgments: An Impossibility Result." *Economics and Philosophy* 18: 89–110.

———. 2004a. "Democracy in Animal Groups: A Political Science Perspective." *Proceedings of the National Academy of Sciences of the United States* 98: 10214–19.

———. 2004b. "On the Significance of the Absolute Margin." *British Journal for the Philosophy of Science* 55, no. 3: 521–44.

———. 2005. "On the Many as One." *Philosophy and Public Affairs*, 33, no. 4: 377–90.

———. 2006. "The Discursive Dilemma and Public Reason." *Ethics* 116: 362–402.

———. 2012. "Collective Wisdom: A Judgment Aggregation Perspective." In *Collective Wisdom: Principles and Mechanisms*, edited by H. Landemore and J. Elster, 203–29. Cambridge: Cambridge University Press.

List, Christian, and Robert E. Goodin. 2001. "Epistemic Democracy: General-izing the Condorcet Jury Theorem." *Journal of Political Philosophy* 9, no. 3: 227–306.

List, Christian, Robert C. Luskin, James Fishkin, and Ian McLean. 2006. "Delib-eration, Single-Peakedness, and the Possibility of Meaningful Democracy: Evidence from Deliberative Polls." PSPE Working Paper No. 1. Available at http://ssrn.com/abstract=1077752.

List, Christian, and Philip Pettit. 2002. "Aggregating Sets of Judgments: An Impossibility Result." *Economics and Philosophy* 18: 89–110.

———. 2005a. "Group Agency and Supervenience." *Southern Journal of Philoso-phy* 44, no. 1: 85–105.

———. 2005b. "On the Many as One." *Philosophy and Public Affairs* 33, no. 4: 377–90.

Lupia, Arthur. 2006. "How Elitism Undermines the Study of Voter Competence." *Critical Review* 18, no. 1–3: 217–32.

Lupia, Arthur, and Mathew D. McCubbins. 1998. *The Democratic Dilemma: Can Citizens Learn What They Need to Know?* Cambridge: Cambridge Uni-versity Press.

Luskin, Robert C. 1987. "Measuring Political Sophistication." *American Journal of Political Science* 31: 856–99.

———. 2002. "From Denial to Extenuation (and Finally Beyond): Political Sophistication and Citizen Performance." In *Thinking about Political Psychol-ogy*, edited by J. H. Kuklinski, 281–305. New York: Cambridge University Press.

Machiavelli, Niccolò. 1996. *Discourses on Livy*. Translated by H. Mansfield and N. Tarcov. Chicago: University of Chicago Press.

———. 1998. *The Prince*. Edited by Q. Skinner and R. Price. Cambridge: Cam-bridge University Press.

———. (1861) 2010. *Considerations on Representative Government*. LaVergne, TN: Book Jungle.

Mackay, Charles. (1841) 1995. *Extraordinary Popular Delusions and the Mad-ness of Crowds*. London: Wordsworth.

Mackie, Gerry. 1998. "All Men Are Liars: Is Democracy Meaningless?" In *Delib-erative Democracy*, edited by J. Elster. 69–96. Cambridge: Cambridge Univer-sity Press.

———. 2003. *Democracy Defended*. Cambridge: Cambridge University Press.

———. 2006. "Does Democratic Deliberation Change Minds?" *Philosophy, Poli-tics and Economics* 5, no. 3: 279–303.

———. 2012. "Rational Ignorance and Beyond." In *Collective Wisdom: Prin-ciples and Mechanisms*, edited by H. Landemore and J. Elster: 290–318. Cam-bridge: Cambridge University Press.

———. 2008. "Why It's Rational to Vote." Unpublished manuscript. Available at http://www.polisci.ucsd.edu/~gmackie/documents/RationalVoting.pdf.

Mandeville, Thomas. 1996. *Understanding Novelty: Information, Technological Change and the Patent System*. Norwood, NJ: Ablex.

Mang Ling Lee, Theresa. 1997. *Politics and Truth: Political Theory and the Post-modernist Challenge*. Albany: State University of New York Press.

Manin, Bernard. 1987. "On Legitimacy and Political Deliberation." *Political Theory* 15, no. 3: 338–68.

———. 1997. *The Principles of Representative Government.* Cambridge: Cambridge University Press.

———. 2005. "Democratic Deliberation: Why We Should Promote Debate rather than Discussion." Paper delivered at the Program in Ethics and Public Affairs Seminar, Princeton University, October 13.

Mann, Michael. 2005. *The Dark Side of Democracy: Explaining Ethnic Cleansing.* Cambridge: Cambridge University Press.

Mansbridge, Jane. 1980. *Beyond Adversary Democracy.* New York: Basic Books.

———. 1992. "A Deliberative Theory of Interest Representation." In *The Politics of Interests: Interest Groups Transformed*, edited by M. P. Petracca, 32–57. Boulder, CO: Westview Press.

———. 1994. "Using Power/Fighting Power." *Constellations* 1, no. 1: 53–73.

———. 1999. "Should Blacks Represent Blacks and Women Represent Women? A Contingent 'Yes.'" *Journal of Politics* 61, no. 3: 628–57.

———. 2003. "Rethinking Representation." *American Political Science Review* 97, no. 4: 515–28.

———. 2006. "Conflict and Self-Interest in Deliberation." In *Deliberative Democracy and Its Discontents: National and Post-national Challenges*, edited by S. Besson and J. L. Martí, 107–32. Burlington, VT: Ashgate.

———. 2010a. "Deliberative Polling as the Gold Standard." Review of *When the People Speak: Deliberative Democracy and Public Consultation*, by James S. Fishkin. *Good Society Journal* 19, no. 1: 55–61.

———. 2010b. "The Place of Self-Interest and the Role of Power in Deliberative Democracy." *Journal of Political Philosophy* 18, no. 1: 64–100.

Mansbridge, Jane, James Bohman, Simone Chambers, David Estlund, Andreas Føllesdal, Archon Fung, Cristina Lafont, Bernard Manin, and José Luis Martí. 2010. "The Place of Self-Interest and the Role of Power in Deliberative Democracy." *Journal of Political Philosophy* 18, no. 1: 64–100.

Margolis, Howard. 2001. "Game Theory and Juries." *Journal of Theoretical Politics* 13, no. 4: 425–35.

Martí, José Luis. 2006. "The Epistemic Conception of Deliberative Democracy Defended." In *Democracy and Its Discontents: National and Post-national Challenges*, edited by S. Besson and J. L. Martí, 27–56. Burlington, VT: Ashgate.

McCormick, John P. 2011. *Machiavellian Democracy.* Cambridge: Cambridge University Press.

McMahon, Christopher. 1997. *Democracy and Authority: A General Theory of Government and Management.* Princeton, NJ: Princeton University Press.

Mendelberg, Tali. 2002. "The Deliberative Citizen: Theory and Evidence." *Political Decision Making, Deliberation and Participation* 6, no. 1: 151–93.

Mercier, Hugo. 2011a. "On the Universality of Argumentative Reasoning." *Journal of Cognition and Culture* 11: 85–113.

———. 2011b. "Reasoning Serves Argumentation in Children." *Cognitive Development* 26, no. 3: 177–91.

———. 2011c. "What Good Is Moral Reasoning?" *Mind and Society* 10, no. 2: 131–48.

———. 2011d. "When Experts Argue: Explaining the Best and the Worst of Reasoning." *Argumentation* 25, no. 3: 313–27.

———. 2011e. "Reasoning Serves Argumentation in Children." *Cognitive Development* (26)3: 177–91.

Mercier, Hugo, and Hélène Landemore. 2012. "Reasoning Is for Arguing: Understanding the Successes and Failures of Deliberation." *Political Psychology* 33, no. 2: 243–58.

Mercier, Hugo, and Dan Sperber. 2009. "Intuitive and Reflective Inferences." In *In Two Minds,* edited by J.S.B.T. Evans and K. Frankish, 149–70. New York: Oxford University Press.

———. 2011a. "Why Do Humans Reason? Arguments for an Argumentative Theory." *Behavioral and Brain Sciences* 34, no. 2, 57–74.

———. 2011b. "Argumentation: Its Adaptiveness and Efficacy." *Behavioral and Brain Sciences* 34, no. 2, 94–111.

Meyer, William J. 1974. "Democracy: Needs over Wants." *Political Theory* 2, no. 2: 197–214.

Michaelsen, Larry K., Warren E. Watson, and Robert H. Black. 1989. "A Realistic Test of Individual versus Group Consensus Decision Making." *Journal of Applied Psychology* 74, no. 5: 834–39.

Michelet, Jules. (1868) 1998. *Histoire de la Révolution française*, vol 2. Paris: Lafont.

Mill, John Stuart. (1861) 2010. *Considerations on Representative Government.* LaVergne, TN: Book Jungle.

———. (1859) 1993b. *On Liberty.* Cambridge: Cambridge University Press.

Millikan, Ruth G. 1987. *Language, Thought, and Other Biological Categories: New Foundations for Realism.* Cambridge, MA: MIT Press.

Moshman, David, and Molly Geil. 1998. "Collaborative Reasoning: Evidence for Collective Rationality." *Thinking and Reasoning* 4, no. 3, 231–48.

Mouffe, Chantal. 1998. "Deliberative Democracy or Agonistic Pluralism?" *Dialogue International Edition* no. 07–08: 9.

———. 2000a. *Deliberative Democracy or Agonistic Pluralism.* Wien: Institut für Höhere Studien.

———. 2000b. *The Democratic Paradox.* London: Verso.

———. 2005. *On the Political.* London: Routledge.

Mueller, Dennis C., Robert D. Tollison, and Thomas D. Willet. 2011. "Representative Democracy via Random Selection." In *Lotteries in Public Life: A Reader,* edited by P. Stone, 47–58. Exeter: Imprint Academic.

Muhlberger, Peter, and Lori M. Weber. 2006. "Lessons from the Virtual Agora Project: The Effects of Agency, Identity, Information, and Deliberation on Political Knowledge." *Journal of Public Deliberation* 2, no. 1: 13.

Mulgan, Richard G. 1984. "Lot as a Democratic Device of Selection." *Review of Politics* 46: 539–60.

———. 2011. "Lot as a Democratic Device of Selection." In *Lotteries in Public Life: A Reader,* edited by P. Stone, 113–32. Exeter: Imprint Academic.

Nagel, Thomas. 1986. *The View from Nowhere.* Oxford: Oxford University Press.

Nelson, William. 1980. *On Justifying Democracy.* Boston: Routledge and Kegan Paul.

Nickerson, Raymond S. 1998. "Confirmation Bias: A Ubiquitous Phenomenon in Many Guises." *Review of General Psychology* 2, no. 2: 175–220.

Niemeyer, Simon. 2004. "Deliberation in the Wilderness: Displacing Symbolic Politics." *Environmental Politics* 13, no. 2: 347–72.

Niemeyer, Simon, and John S. Dryzek. 2007. "The Ends of Deliberation: Meta-Consensus and Inter-Subjective Rationality as Ideal Outcomes." *Swiss Political Science Review* 13, no. 4: 497–526.

Nietzsche, Friedrich. 1989. *Beyond Good and Evil: Prelude to a Philosophy of the Future*. Translated by Walter Kaufman. New York: Random House.

Nino, Carlos S. 1996. *The Constitution of Deliberative Democracy*. New Haven, CT: Yale University Press.

Nocera, Joe. 2006. "The Future Divined by the Crowd. *New York Times*. March 11 (with a correction published on March 17).

Norman, Donald A. 1991. "Cognitive Artifacts." In *Designing Interaction: Psychology at the Human-Computer Interface,* edited by J. M. Carroll, 17–38. New York: Cambridge University Press.

Noveck, Beth Simone. 2009. *Wiki-Government: How Technologies Can Make Government Better, Democracy Stronger, and Citizens More Powerful*. Washington: Brookings Institution Press.

O'Leary, Kevin. 2006. *Saving Democracy: A Plan for Real Representation in America*. Palo Alto, CA: Stanford Law Books.

Ober, Josiah. 2010. *Democracy and Knowledge: Innovation and Learning in Classical Athens*. Princeton, NJ: Princeton University Press.

———. 2012. "Epistemic Democracy in Classical Athens: Sophistication, Diversity, and Innovation." In *Collective Wisdom: Principles and Mechanisms*, edited by H. Landemore and J. Elster, 118–47. Cambridge: Cambridge University Press.

Page, Benjamin, and Marshall M. Bouton. 2006. *The Foreign Policy Disconnect. What Americans Want from Our Leaders but Don't Get*. Chicago: University of Chicago Press.

Page, Benjamin I., and Robert Y. Shapiro. 1992. *The Rational Public: Fifty Years of Trends in Americans' Policy Preferences*. Chicago: University of Chicago Press.

Page, Scott E. 2007. *The Difference: How the Power of Diversity Creates Better Groups, Firms, Schools, and Societies*. Princeton, NJ: Princeton University Press.

Panning, William H. 1986. "Information Pooling and Group Decisions in Nonexperimental Settings." In *Information Pooling and Group Decision Making*, edited by Bernard Grofman and Guillermo Owen. Greenwich, CT: JAI Press.

Pateman, Carole. 1970. *Participation and Democratic Theory*. Cambridge: Cambridge University Press.

Peter, Fabienne. 2009. *Democratic Legitimacy*. New York: Routledge.

Pettit, Philip. 2001. "Deliberative Democracy and the Discursive Dilemma." *Philosophical Issues* 11: 268–99.

———. 2003. "Deliberative Democracy, the Discursive Dilemma, and Republican Theory." In *Debating Deliberative Democracy*, edited by J. S. Fishkin and P. Laslett, 138–62. Oxford: Blackwell.

———. 2004. "Groups with a Mind of Their Own." In *Socializing Metaphysics*, edited by F. Schmitt, 167–93. New York: Rowman and Littlefield.

Philonenko, Alexis. 1984. *Jean-Jacques Rousseau et la pensée du malheur*. Paris: Vrin.

Pitkin, Hanna. 1967. *The Concept of Representation*. Berkeley: University of California Press.

Plato. *Protagoras*. Translated by Benjamin Jowett. E-book available at http://classics.mit.edu/Plato/protagoras.html.

Plott, Charles R., and Kay-Yut Chen. 2002. "Information Aggregation Mechanisms: Concept, Design and Implementation for a Sales Forecasting Problem." Social Science Working Paper 1131, California Institute of Technology. Available at http://www.hss.caltech.edu/SSPapers/wp1131.pdf.

Popkin, Samuel L. 1994. *The Reasoning Voter: Communication and Persuasion in Presidential Campaigns*. 2nd ed. Chicago: University of Chicago Press.

Popkin, Samuel L., and Michael Dimock. 1999. "Political Knowledge and Citizen Competence." In *Citizen Competence and Democratic Institutions*, edited by S. Elkin and K. Soltan. University Park: Pennsylvania State University Press.

Posner, Richard A. 2002. "Dewey and Democracy." *Transactional Viewpoints* 1, no. 3: 1–4.

———. 2003. *Law, Pragmatism, and Democracy*. Cambridge, MA: Harvard University Press.

———. 2004. "Law, Pragmatism, and Democracy: Reply to Somin." *Critical Review* 16, no. 4: 463–69.

Przeworski, Adam. 1999. "Minimalist Conception of Democracy: A Defense." In *Democracy's Value*, edited by I. Shapiro and C. Hacker-Cordon, 23–55. Cambridge: Cambridge University Press.

———. 2000. *Democracy and Development: Political Institutions and Well-Being in the World*. Cambridge: Cambridge University Press.

Putnam, Hilary. 2003. *The Collapse of the Fact/Value Dichotomy and Other Essays*. Cambridge, MA: Harvard University Press.

Rancière, Jacques. 2005. *La haine de la démocratie*. Paris: La Fabrique.

Rawls, John. 1980. "Kantian Constructivism in Moral Theory." *Journal of Moral Philosophy* 77, no. 9: 515–72.

———. 1993. *Political Liberalism*. New York: Columbia University Press.

———. 1995. "Political Liberalism: Reply to Habermas." *Journal of Philosophy* 92, no. 3: 132–80.

———. (1971) 1999. *Theory of Justice*. Cambridge, MA: Belknap Press.

Raz, Joseph. 1979. *The Authority of Law: Mssays on Law and Morality*. Oxford: Clarendon Press and New York: Oxford University Press.

———. 1990. "Facing Diversity: The Case of Epistemic Abstinence." *Philosophy and Public Affairs* 19, no. 1: 3–46.

Rehfeld, Andrew. 2005. *The Concept of Constituency: Political Representation, Democratic Legitimacy and Institutional Design*. Cambridge: Cambridge University Press.

———. 2006. "Towards a General Theory of Political Representation." *Journal of Politics* 68, no. 1.

———. 2008. "Extremism in the Defense of Moderation: A Response to My Critics." *Polity* 40: 254–71.

———. 2009. "Representation Rethought." *American Political Science Review* 103, no. 2: 214–30.

———. 2010. "On Quotas and Qualifications for Office." In *Political Representation*, edited by Ian Shapiro, Susan Stokes, Elisabeth Woods, and Alexander Kirschner, 236–68. New York: Cambridge University Press.

Rheingold, Howard. 2003. *Smart Mobs*. Cambridge, MA: Perseus Publishing.

Riker, William H. 1982. *Liberalism against Populism: A Confrontation between the Theory of Democracy and the Theory of Social Choice*. San Francisco: W. H. Freeman.

Risse, Mathias. 2001. "The Virtuous Group—Foundations for the Argument from the Wisdom of the Multitude." *Canadian Journal of Philosophy* 31, 38–85.

———. 2004. "Arguing for Majority Rule." *Journal of Political Philosophy* 12, no. 1: 41–64.

Robson, John M. 1968. *The Improvement of Mankind: The Social and Political Thought of John Stuart Mill*. London: Routledge.

Rorty, Richard. 1979. *Philosophy and the Mirror of Nature*. Princeton, NJ: Princeton University Press.

———. 2006. *Take Care of Freedom and Truth Will Take Care of Itself: Interviews with Richard Rorty*. Stanford, CA: Stanford University Press.

Rosanvallon, Pierre. 1993. "Histoire du mot démocratie à l'époque moderne." *La Pensée Politique* 1: 11–29.

Ross, Lee, Mark R. Lepper, and Michael Hubbard. 1975. "Perseverance in Self-perception and Social Perception: Biased Attributional Processes in the Debriefing Paradigm." *Journal of Personality and Social Psychology* 32: 880–92.

Rousseau, Jean-Jacques. (1762) 1997. *The Social Contract and Other Later Political Writings*. Edited and translated by V. Gourevitch. Cambridge: Cambridge University Press.

Rovane, Carol. 1998. *The Bounds of Agency: An Essay in Revisionary Metaphysics*. Princeton, NJ: Princeton University Press.

Rumelhart, David E., Paul Smolensky, James L. McClelland, and Geoffrey Hinton. 1986. "Schemata and Sequential Thought Processes in PDP Models." In *Parallel Distributed Processing: Explorations in the Microstructure of Cognition*. Vol. 2, *Psychological and Biological Models*, edited by J. L. McClelland and D. Rumelhart, 7–57. Cambridge, MA: MIT Press.

Ryfe, David M. 2005. "Does Deliberative Democracy Work?" *Annual Review of Political Science* 8: 49–71.

Sabbagh, Daniel. 2003. *L'Égalité par le droit: Les paradoxes de la discrimination positive aux États-Unis*. Paris: Economica.

Salovey, Peter, and John D. Mayer. (1990) 1998. "Emotional Intelligence." In *Human Emotions: A Reader*, edited by K. Oatley, J. M. Jenkins, and N. L. Stein, 313–20. Oxford: Blackwell Publishers.

Sanders, Lynn. 1997. "Against Deliberation." *Political Theory* 25, no. 3: 347–76.

Scanlon, Thomas. 1998. *What We Owe to Each Other*. Cambridge, MA: Belknap Press.

Schauer, Frederick. 1999. "Talking as a Decision Procedure." In *Deliberative Politics: Essays on Democracy and Disagreement*, edited by S. Macedo, 17–27. Oxford: Oxford University Press.

Schlozman, Kay Lehman, Sidney Verba, and Henry E. Brady. 1999. "Civic Participation and the Equality Problem." In *Civic Engagement and American Democracy*, edited by T. Skocpol and M. P. Fiorina. Washington, DC: Brookings Institution Press.

Schofield, Norman. 2002. "Madison and the Founding of the Two-Party System." In *James Madison: The Theory and Practice of Republican Government*, edited by S. Kernell, 302–28. Stanford, CA: Stanford University Press.

Schumpeter, Joseph. (1942) 1975. *Capitalism, Socialism, and Democracy*. New York: Harper and Brothers.

Schwartzberg, Melissa. 2009. "Shouts, Murmurs, and Votes: Acclamation and Aggregation in Ancient Greece." *Journal of Political Philosophy* 18, no. 4: 448–68.

Sen, Amartya. 1970. *Collective Choice and Social Welfare*. San Francisco: Holden-Day.

———. 1999. *Democracy as Development*. New York: Anchor Books.

Servan-Schreiber, Emile. 2012. "Prediction Markets: Trading Uncertainty for Collective Wisdom." In *Collective Wisdom: Principles and Mechanisms*, edited by H. Landemore and J. Elster, 21–37. Cambridge: Cambridge University Press.

Shafir, Eldar, Itamar Simonson, and Amos Tversky. 1993. "Reason-Based Choice." *Cognition* 49, no. 1–2: 11–36.

Shapiro, Ian. 1999. "Enough of Deliberation: Politics Is About Interest and Power." In *Deliberative Politics: Essays on Democracy and Disagreement*, edited by S. Macedo, 28–38. New York: Oxford University Press.

———. 2003a. *The Moral Foundations of Politics*. New Haven, CT: Yale University Press.

———. 2003b. *The State of Democratic Theory*. Princeton, NJ: Princeton University Press.

Simmons, John A. 2001. *Justification and Legitimacy: Essays on Rights and Obligations*. Cambridge: Cambridge University Press.

Simon, Herbert. 1957. *Models of Man: Social and Rational: Mathematical Essays on Rational Human Behavior in a Social Setting*. New York: Wiley.

Sintomer, Yves. 2007. *Le pouvoir au peuple: Jury citoyens, tirage au sort, et démocratie participative*. Paris: La Découverte.

Skinner, Quentin. 1973. "The Empirical Theorists of Democracy and Their Critics: A Plague on Both Their Houses." *Political Theory* 1, no. 3: 287–306.

Skorupski, John, ed. 1998. *The Cambridge Companion to Mill*. Cambridge: Cambridge University Press.

Slavin, Robert E. 1996. "Research on Cooperative Learning and Achievement: What We Know, What We Need to Know." *Contemporary Educational Psychology* 21, no. 1: 43–69.

Sniezek, Janet, and Rebecca Henry. 1989. "Accuracy and Confidence in Group Judgment." *Organizational Behavior and Human Decision Processes* 43: 1–28.

Somin, Ilya. 1998. "Voter Ignorance and the Democratic Ideal." *Critical Review* 12, no. 4: 413–58.

———. 1999. "Resolving the Democratic Dilemma?" *Yale Journal on Regulation* 16: 401–14.

———. 2004a. "Political Ignorance and the Countermajoritarian Difficulty: A New Perspective on the Central Obsession of Constitutional Theory." *Iowa Law Review* 89: 1287–1371.

———. 2004b. "Richard Posner's Democratic Pragmatism." *Critical Review* 16, no. 1: 1–21.

Sperber, Dan. 2000. "Metarepresentations in an Evolutionary Perspective." In *Metarepresentations: A Multidisciplinary Perspective*, edited by D. Sperber, 117–37. Oxford: Oxford University Press.

———. 2001. "An Evolutionary Perspective on Testimony and Argumentation." *Philosophical Topics*, 29: 401–13.

Spinoza, Benedict de. 1670. *Tractatus Theologico-Politicus*. Available at http://www.worldwideschool.org/library/books/relg/christiantheology/ATheologico-PoliticalTreatise/chap16.html.

Stanovich, K. 2004. *The Robot's Rebellion*. Chicago: University of Chicago Press.

Steiner, Jürg 2008. "Concept Stretching: The Case of Deliberation." *European Political Science* 7: 186–90.

Steiner, Jürg, and Robert H. Dorff. 1980. "Decision by Interpretation: A New Concept for an Often Overlooked Decision Mode." *British Journal of Political Science* 11, no. 1: 1–13.

Sternberg, Robert. J. 1985. *Beyond IQ: A Triarchic Theory of Human Intelligence*. New York: Cambridge University Press.

Stone, Peter. 2007. "Why Lotteries Are Just." *Journal of Political Philosophy* 15, no. 3: 276–95.

———. 2009. "The Logic of Random Selection." *Political Theory* 37: 375–97.

———. 2010. *The Luck of the Draw: The Role of Lotteries in Decision-Making*. Oxford: Oxford University Press.

———, ed. 2011. *Lotteries in Public Life: A Reader*. Exeter: Imprint Academic.

Sullivan, Eileen P. 1998. "Liberalism and Imperialism: J. S. Mill's Defense of the British Empire." In *John Stuart Mill's Social and Political Thought: Critical Assessments*, edited by G. W. Smith. New York: Routledge.

Sunstein, Cass. 1988. "Beyond the Republican Revival." *Yale Law Journal* 97: 1539–90.

———. 2002. "The Law of Group Polarization." *Journal of Political Philosophy* 10, no. 2: 175–95.

———. 2003. *Why Societies Need Dissent*. Cambridge, MA: Harvard University Press.

———. 2006. *Infotopia: How Many Minds Produce Knowledge*. London: Oxford University Press.

———. 2007. "Deliberating Groups versus Prediction Markets (or Hayek's Challenge to Habermas)." *Episteme: Journal of Social Epistemology* 3: 192–213.

Surowiecki, James. 2004. *The Wisdom of Crowds: Why the Many Are Smarter than the Few and How Collective Wisdom Shapes Business, Economies, Societies, and Nations*. New York: Doubleday.

Suskind, Ron. 2011. *Confidence Men: Washington, Wall Street, and the Education of a President*. New York: HarperCollins.

Talisse, Robert. 2004. "Does Public Ignorance Defeat Deliberative Democracy?" *Critical Review* 10, no. 4: 455–63.

———. 2009. *Democracy and Moral Conflict*. Cambridge: Cambridge University Press.

Tangian, Andranik S. 2000. "Unlikelihood of Condorcet's Paradox in a Large Society." *Social Choice and Welfare* 17: 337–65.

Tetlock, Philip. 2005. *Expert Political Judgment: How Good Is It? How Can We Know?* Princeton, NJ: Princeton University Press.

Thompson, Dennis. 2008. "Deliberative Democratic Theory and Empirical Political Science." *Annual Review of Political Science* 11: 497–520.

Tocqueville, Alexis de. 2002. *Democracy in America*. Translated by H. Mansfield and D. Winthrop. Chicago: University of Chicago Press.

Tuck, Richard. 2008. *Free Riding*. Cambridge, MA: Harvard University Press.

Urbinati, Nadia. 2002. *Mill on Democracy: From the Athenian Polis to Representative Government*. Chicago: University of Chicago Press.

———. 2006. *Representative Democracy: Principles and Genealogy*. Chicago: University of Chicago Press.

Urfalino, Philippe. 2007. "La décision par consensus apparent: Nature et propriétés." *Revue européenne des sciences sociales* 45, no. 136: 34–59. An earlier version of this text is available in English at http://cesta.ehess.fr/document.php?id=126.

———. 2010. "Deciding as Bringing Deliberation to a Close." *Social Science Information* 49, no. 1 (special issue, "Rules of Collective Decision"): 109–38.

———. 2011. "Decision by Non-Opposition: Silence Means Consent . . . but Not Necessarily Approval." Paper presented at the Yale Political Theory Workshop, New Haven, CT, April 8.

———. 2012. "Sanior Pars and Major Pars in Contemporary Areopagus: Medicine Evaluation Committees in France and United-States." In *Collective Wisdom: Principles and Mechanisms*, edited by H. Landemore and J. Elster, 173–202. Cambridge: Cambridge University Press.

Urken, Arnold. 1991. "The Condorcet-Jefferson Connection and the Origins of Social Choice Theory." *Public Choice* 72: 213–36.

Urken, Arnold B., and Stephen Traflet. 1984. "Optimal Jury Design." *Jurimetrics* 24: 218–35.

Waldron, Jeremy. 1989. "Democratic Theory and the Public Interest: Condorcet and Rousseau Revisited." *American Political Science Review* 83, no. 4: 1317–40.

———. 1993. "Rights and Majorities: Rousseau Revisited." In J. Waldron, *Liberal Rights: Collected Papers, 1981–1991*. Cambridge: Cambridge University Press: 392–421.

———. 1995. "The Wisdom of the Multitude: Some Reflections on Book 3, Chapter 11 of Aristotle's *Politics*." *Political Theory* 23, no. 4: 563–84.

———. 1999. *The Dignity of Legislation*. Cambridge: Cambridge University Press.

———. 2001. *Law and Disagreement*. Oxford: Oxford University Press.

Walzer, Michael. 1981. *Spheres of Justice: A Defense of Pluralism and Equality*. New York: Basic Books.

Warren, Mark E. 2009. "Two Trust-Based Uses of Mini-Publics in Democracy." Prepared for Conference on Democracy and the Deliberative Society, University of York.

Warren, Mark E., and Hilary Pearse. 2008. *Designing Deliberative Democracy: The British Columbia Citizens' Assembly.* Cambridge: Cambridge University Press.

Wason, Peter C. 1966. "Reasoning." In B. M. Foss, *New Horizons in Psychology.* Harmondsworth, UK: Penguin.

Wason, Peter C., and Philip G. Brooks. 1979. "THOG: The Anatomy of a Problem." *Psychological Research* 41, no. 1: 79–90.

Wertsch, James V. 1998. *Mind as Action.* New York: Oxford University Press.

White, Stephen K. 1995. *The Cambridge Companion to Habermas.* Cambridge: Cambridge University Press.

Wilentz, Sean. 2005. *The Rise of American Democracy.* New York: Norton.

Williams, Bernard. 1985. *Ethics and the Limits of Philosophy.* Cambridge, MA: Harvard University.

Wilson, Timothy D., and Jonathan W. Schooler. 1991. "Thinking Too Much: Introspection Can Reduce the Quality of Preferences and Decisions." *Journal of Personality and Social Psychology* 60: 181–92.

Wingo, Ajume H. 2005. "Modes of Public Reasoning in the Islamic/West Debate." Unpublished paper.

Wolfers, Justin, and Eric Zitzewitz. 2004. "Prediction Markets." *Journal of Economic Perspectives* 18, no. 2: 107–26.

Young, H. Peyton. 1988. "Condorcet's Theory of Voting." *American Political Science Review* 82, no. 4: 1231–44.

Young, Iris Marion. 2000. *Inclusion and Democracy.* Oxford: Oxford University Press.

Zaller, John R. 1992. *The Nature and Origins of Mass Opinion.* New York: Cambridge University Press.

Index

Aborigines (Australia), 113n26
Academy Awards, predictions of winners, 175n30
Ackerman, Bruce, 34, 105, 180
Adams, John, 108
affirmative action, 103, 139. *See also* quotas and gerrymandering
agenda setting: potential manipulation in, 155–56, 190n8, 237; random mechanism for, 190
aggregation: logic of cognitive diversity in (tables and explanations), 169–72; market-based procedure for (collective wisdom), 173–84. *See also* binary choices; Condorcet Jury Theorem; information markets; law of large numbers; majority rule; Miracle of Aggregation; voting
agonistic pluralists, 95, 224, 225
Althaus, Scott, 200–201
AmericaSpeaks, 133
Anderson, Elizabeth, 48–49, 148
antiauthoritarian objection: Arendt's view of truth and politics as, 224–27; context of, 223–24; Rawls's epistemic abstinence and, 44n25, 227–30
antiforeigner bias, 33n10, 196b14
antimarket bias, 33, 165, 196, 198
apathy of citizens, 33–36
Aquinas, Thomas, 63
Arendt, Hannah: on historical facts, 214; on truth and politics, 209, 222, 223–27
argumentative theory of reasoning: approach to, 119, 123–24; argument production vs. assessment in, 126–27; classical view vs., 124–25;

definition of, 125–30; deliberation within vs. with others and, 130–36, 138; development of, 24–25; group polarization idea and, 118, 124, 136–42; setting of, 127; summary of, 143–44
Aristotle: on collective wisdom, 2, 59–64; on deliberation, 91; on democracy, 27; students of, 63–64; work: *Politics*, 59–64
Arrow, Kenneth J., 174. *See also* Arrow's Impossibility Theorem
Arrow's Impossibility Theorem (1953): application of, 39n23; approach to, 25; basics of, 39, 149; irrelevance to epistemic argument, 40, 189–90; majority rule objections in, 186–89
artificial intelligence, 20n17
Asad, Muhammad, 54–55
Athenian democracy: epistemic analysis of, 2–3, 44, 49, 61, 238; equal right of speech defended in, 13, 58; *isegoria* principle of, 56, 58–59, 112n24, 203; myth underlying practices of, 57–58; ostracism practice in, 183; Sicily expedition and, xvi, 211–12, 239
Austen-Smith, David, 155n11
Australia: citizens' jury in, 132–36; deliberative poll in, 113n26

Baker, Keith M., 151n7
Banks, Jeffrey S., 155n11
bargaining, 92n3, 93, 94, 191
base-rate neglect (or fallacy), 205
Beitz, Charles R., 50–51, 210n29
Berlin, Isaiah, 206, 222